Topological Insulators
Materials and Applications

Edited by

Inamuddin[1], Tariq Altalhi[2], Mohammad Abu Jafar Mazumder[3,4]

[1]Department of Applied Chemistry, Zakir Husain College of Engineering and Technology, Faculty of Engineering and Technology, Aligarh Muslim University, Aligarh-202002, India

[2]Department of Chemistry, College of Science, Taif University, 21944 Taif, Saudi Arabia

[3]Chemistry Department, King Fahd University of Petroleum & Minerals, Dhahran 31261, Saudi Arabia

[4]Interdisciplinary Research Center for Advanced Materials, King Fahd University of Petroleum & Minerals, Dhahran 31261, Saudi Arabia

Published by **Materials Research Forum LLC**
Millersville, PA 17551, USA

Published as part of the book series
Materials Research Foundations
Volume 154 (2024)
ISSN 2471-8890 (Print)
ISSN 2471-8904 (Online)

Print ISBN 978-1-64490-284-4
eBook ISBN 978-1-64490-285-1

Distributed worldwide by

Materials Research Forum LLC
105 Springdale Lane
Millersville, PA 17551
USA
https://www.mrforum.com

Manufactured in the United States of America
10 9 8 7 6 5 4 3 2 1

Table of Contents

Preface

A topological insulator is an area that has yet to be fully explored and developed. The charge-induced bandgap fluctuation in the best-known bismuth-chalcogenide-based topological insulators is approximately 10MeV in magnitude. The major focus of the study has shifted to the investigation of the presence of high-symmetry electronic bands as well as the utilization of easily produced materials. Half-Heusler compounds are one of the possibilities. These crystal forms can contain a wide variety of elements. Eighteen electrons half Heusler compound with a non-trivial band structure which exhibits an order similar to the well-acquainted topological insulator materials was predicted using calculations by the first principle, to have a band structure with band order similar to the popular 2D and 3D topological insulator materials. When tested in real-world tests, these groups of materials have not yet come up with any evidence of inherent topological insulator behavior.

Theory for the 2D topological insulator has been presented in the past decade with the most recent being published in 2008. Following that, in quantum wells of mercury-telluride packed in between cadmium-telluride, the theoretical prediction for a 2D topological insulator with 1D helical-edge states was achieved in quantum-wells of cadmium-telluride. In subsequent tests, it was discovered that the transport caused by 1D helical edge states was occurring. No one anticipated that 3D topological insulators would be discovered in the binary compounds containing bismuth until 2007. In particular, no one could predict that "strong topological insulators" would prevail to exist that could not be brought to multiple replicas of the Quantum-Spin-Hall-state until 2007. As the subject of topological insulators is still in the nascent stage, there is growing research and knowledge in the emerging field. This book is intended to provide the readers with an understanding of the need and application of these materials. There are nine chapters in this book summarized as given below:

Chapter 1 details the basic terminologies of topological insulators having conducting edge states. They exhibit unique electronic properties that arise from topological order, time-reversal symmetry, and the presence of protected edge states. Spin-orbit interaction and topological invariants also played significant roles in understanding as well as characterizing these materials.

Chapter 2 entails a brief overview and history of 1D topological insulators. Comprehensive details on properties of these topological insulators such as quantum spin hall effect, spin-polarized electrons, and low power dispersion, classification based on symmetry, dimension, and parity of Dirac points different synthesis techniques along with generations of topological insulators, their advanced models, topological insulators in 10-fold and future prospectives are discussed.

Chapter 3 details the overview of the history of making the topographical insulator and the knowledge acquired from the past. It also discusses the new emerging field of topographical quantum and its architecture when combined with superconductors.

Chapter 4 deliberates magnetic topological insulators, the origin of magnetism in them, and their types such as intrinsic magnetic insulators. Experimental observation of magnetism, antiferromagnetic phase, ferromagnetic phase, quantum hall effect, quantum anomalous hall effect, and quantum spin hall effect in the intrinsic magnetic TIs have been deliberated.

Chapter 5 explores a class of superconductors called topological superconductors. The background theory, the emergence of Majorana fermions, properties such as spin current, anomalous Josephson effect, and nematicity of topological superconductors along with potential superconductivity candidates in these topological semiconductors such as tin-based, iron-based topological superconductors are briefly discussed.

Chapter 6 discusses in detail manganese-doped topological insulators with emphasis on structure and properties.

Chapter 7 offers an overview of topological insulators (TIs) integrated inside a thin-film solar cell, the working bandwidth of photonic crystal, and topological beam splitters. A detailed experimental discussion has been provided about two important 3D TIs i.e. bismuth telluride and bismuth solenoid. Photo-induced structured waves, dynamic optical behavior, and saturable absorbers are also discussed.

Chapter 8 provides recent advances in the field of topological insulators-based ultrafast mode-lock fiber lasers. Mode-locking operation and optical characteristics of Yb-doped and Er-doped mode-locked fiber lasers are experimentally discussed in detail using topological insulators Bi2Te3 and graphene as saturable absorbers, as well its potential uses are also examined.

Chapter 9 provides insights in brief on the topological insulators concerning historical evolution. A pedagogical account informs the readers about various single crystal growth techniques to realize superior-grade topological insulators for consequent roles in customized applications such as quantum technology, catalysis, spintronics, microelectronics, and opto-magnetic devices.

Editors

Inamuddin, Tariq Altalhi, Mohammad Abu Jafar Mazumder

Topological Insulators: Materials and Applications
Materials Research Foundations 154 (2024) 1-20

Materials Research Forum LLC
https://doi.org/10.21741/9781644902851-1

Chapter 1

Fundamental Concepts of Topological Insulators

Remsha Shakeel, Haq Nawaz Bhatti and Amina Khan*

Department of Chemistry, University of Agriculture, Faisalabad, Punjab, Pakistan

aminakhan1649@gmail.com

Abstract

The notion of topological insulators was first introduced to explain the concept of Quantum Hall Effect. The Quantum Hall State (QHS) does not disrupt symmetries but showed fundamental properties (like quantized Hall conductivity, the number of conducting edge-mode) that are not affected by smooth changes in different material parameters and are not subject to change if the system goes through the quantum phase-transition. A topological insulator (TI) just like an ordinary insulator has a large energy gap that is separating the highest-filled electronic band from the lowest empty-band. However, a topological insulator's surface must have gapless electronic states which are protected by the time-reversal symmetry (TRS). Like QHS, having distinctive gapless chiral edge-states on the surface or the edge-states of the topological insulators (TIs) are topologically shielded and reveal conducting states having properties that are unlike any other known 1D and 2D electronic systems. Strong spin-orbit interactions under the conservation of time-reversal symmetry (TRS) are the driving force behind these substances. Moreover, Topological insulators (TIs) were revealed experimentally for the first time in 2007 by the consideration of the condensed-matter physics community which become fully focused on a novel category of materials. The 3D topological insulator's new qualities could result in some fascinating applications because they are very common semiconductors and their topological properties can withstand high temperatures. Hence, Topological insulators (TIs) are those materials that are electrically inert in bulk but can carry out electricity due to their topologically protected electronic edge-state as well as surface states.

Keywords

Quantum Hall, Insulators, Quantized, Topological, Spin

Topological Insulators: Materials and Applications
Materials Research Forum LLC
Materials Research Foundations 154 (2024) 1-20
https://doi.org/10.21741/9781644902851-1

Contents

1.   Introduction

Topological insulator (TI), an emerging semiconductor and quantum material, has a small bulk band gap and a gapless surface state [1]. Topological insulators (TIs) are a unique class of compounds that display distinctive electronic properties and are of great interest in the field of condensed matter physics. They are characterized by their ability to conduct electricity on their surface or edge while remaining insulating in the bulk. The concept of TIs is rooted in the principles of topology, the branch of mathematics which deals with the properties of geometric objects that remain unchanged under ongoing deformations [2]. In the context of materials, topological insulators possess non-trivial topological properties in their electronic band structure.

In a conventional insulator, the electronic band structure consists of a completely occupied valence band and a vacant conduction band, with a distinct energy gap between them. In contrast, topological insulators have a band structure that exhibits nontrivial topology; resulting in the formation of topologically protected edge-states or surfaces within the band gap [3]. These topologically protected states are robust against impurities, disorder, and most external perturbations due to symmetry constraints and the presence of certain topological invariants. They behave as metallic states on the surface or edges of the material while the bulk remains insulating. This behavior is attributed to the protection provided by symmetries such as time-reversal symmetry.

The surface or edge states in topological insulators (TIs) have distinctive properties. They can exhibit spin-momentum locking, from where the electron spin orientation is locked to its momentum direction [4]. This property enables the efficient manipulation and transport of spin information, making topological insulators potentially useful for spintronics and quantum computing applications. Topological insulators can be found in various material systems, including compounds such as bismuth-selenide (Bi_2Se_3), bismuth-telluride (Bi_2Te_3), and antimony-telluride (Sb_2Te_3) [5]. The discovery and understanding of topological insulators have led to significant advancements in material synthesis, characterization techniques, and theoretical understanding. In summary, topological insulators are materials that possess nontrivial topological properties in their electronic band structure. They exhibit metallic states on the surface or edge while being insulated in the bulk. These unique electronic properties make topological insulators promising for various applications in electronics, spintronics, and quantum computing.

The theoretical foundation for the topological insulators began with the invention of the Quantum Hall (QH) effect in 1980, for that Klaus von Klitzing was rewarded with the Nobel Prize in Physics [6]. The quantum Hall (QH) effect revealed the existence of topologically protected edge states in two-dimensional (2D) electronic systems exposed to powerful magnetic fields (MF). In the 1990s, researchers began exploring the possibility of topologically nontrivial states in systems without external MF. In 1998, Charles Kane and Eugene Mele theoretically proposed the concept of Quantum Spin Hall insulators, which predicted the existence of topologically protected states in two-dimensional systems due to the interplay of spin-orbit coupling as well as band topology.

The experimental realization of topological insulators took a significant leap forward in 2005 when a team led by M. Zahid Hasan and Charles Kane discovered the first three-dimensional topological insulator, bismuth antimony ($Bi_{1-x}Sb_x$) alloys. They observed metallic surface states with Dirac-like dispersion, indicative of a topologically protected state [7]. In 2007, a breakthrough in topological insulator research came with the innovation of a new family of materials called ternary chalcogenides, such as bismuth-telluride (Bi_2Te_3) and bismuth-selenide (Bi_2Se_3). A research team led by Shoucheng Zhang and Qi-Kun Xue experimentally confirmed the existence of robust surface states with Dirac-like dispersion in these materials, solidifying the concept of TIs.

The exploration of topological insulators continued to progress in the 2010s, leading to the innovation of new materials, novel phenomena, and the realization of applications. The study of topological insulators expanded into other areas, such as topological superconductors and topological semimetals, further enriching the field [8].

The development of topological insulators has led to significant advancements in materials science, condensed matter physics, and topological quantum phenomena. It has opened up new platforms for research and potential technological breakthroughs in fields such as electronics, spintronics, and quantum computing.

2.   Basic concepts

2.1   Quantum Hall to Quantum Spin Hall

In the present quantum world, atoms and their electrons are able to produce diverse matter states, like crystalline solids, superconductors, and magnets. For the above-mentioned examples, rotational, translational, or gauge symmetries are the symmetries that are used to classify these different states. Before 1980, the broken symmetry principle is used to classify all matter states in a condensed-matter system. The Quantum Hall State (QHS), revealed in 1980 provided the earliest evidence of the quantum state which is devoid of spontaneously broken symmetry. Its behavior is mainly governed by its topology but not by the certain geometry as it was topologically different from all earlier known matter states [9].

A two-dimensional (2D) electron gas in the semiconductor experiences QH effect as a powerful magnetic field (MF) is applied to it. The two counter-flows of electrons are radially divided into distinct 'lanes' present at the top as well as bottom edges of the semiconductor sample in the presence of MF and low temperature. Electrons only move along the semiconductor's edge under these conditions. Only half as many degrees of freedom are positioned at the upper edge of the QH bar as compared to the one-dimensional system with electrons moving in both directions. The QH effect is topologically resilient because of this particular spatial separation, which is symbolized in figure 2a by the symbolic formula "2 = 1 [forward mover] + 1 [backward mover]". Since there is no mechanism for an edge-state electron to turn around when it comes into contact with an impurity as it simply undergoes a diversion and continues moving in a similar direction [10].

In a real one-dimensional system, as depicted in Figure 2b, four channels are created by forward along with backward moving channels for the spin-up as well as a spin-down electron, respectively. Without a magnetic field, it is possible to divide the electron traffic lanes in a TR-invariant way, as shown in the figure by the equation "4 = 2 + 2." We can transfer the remaining two channels toward the bottom-edge while leaving the spin-up forward mover and spin-down backward mover on the top edge. A system that contains these edge states is considered in the QSH state because similar to separated charge transport in the QH state, there is an overall transport of spin forward across the top edge as well as backward across the bottom edge [11]. Although the QSH edge has both forward as well as backward movers, back-scattering by non-magnetic impurities is prohibited.

Only the scenario of single pairings of the QSH edge-states is illustrated by the physical illustration above. Electrons can scatter from the forward to the backward moving channels without changing their spin or causing effective destructive interference, if there are two backward movers and two forward movers in the given system, as in the unseparated one-dimensional system depicted in Figure 2. This leads to dissipation. Hence, the edge states must have an odd number of backward movers or an odd number of forward movers for QSH state to be resilient. The core of the QSH state is the even-odd effect, which is represented by the so-called Z_2 topological quantum number. For this reason, a QSH insulator is also known as a topological insulator (TI). Quantum Hall states (QH) and Spin Quantum Hall states (SQH) are two distinct phenomena observed in certain topological insulators under the influence of a strong magnetic field. While both phenomena involve quantized Hall conductivity, they arise from different mechanisms and exhibit different characteristics.

Quantum Hall states are observed in two-dimensional electronic systems subjected to the strong perpendicular MF. They are characterized by the quantization of Hall-conductivity to values that are integer multiples of a fundamental constant, known as the von Klitzing constant. This quantization is a manifestation of the topological characteristics of the electron system [12].

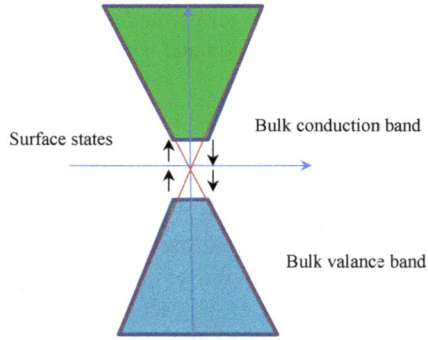

Figure 1. Electronic band structure of the topological insulator

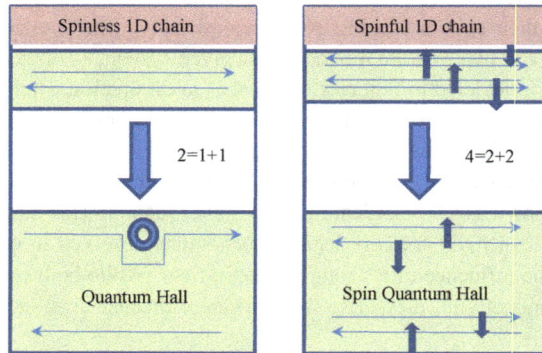

Figure 2. Special separation in (a) spinless 1D system (b) spinful 1D system.

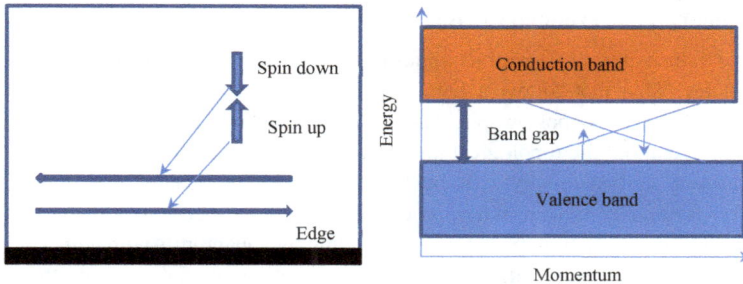

Figure 3. Time reversal symmetry in topological insulator.

In Quantum Hall states, the electrons form two-dimensional electron-gas (2DEG) in which Landau levels, discrete energy levels, are formed due to the magnetic field quantization. These Landau levels are separated by a gap, and the current flow is carried by the edge states of the system [13]. The edge states exhibit a chiral nature, propagating in one direction along the edges while being localized near the sample boundaries. These edge states are topologically protected and tough against any disorder. The quantized Hall conductivity arises from topological distinct properties of the electron system, specifically the presence of a nontrivial Berry phase related to the electron motion in the magnetic field. This Berry phase results in the formation of a bandgap in the bulk of the material, while the edge states remain gapless and conductive. Spin quantum Hall states, also referred to as the Spin Hall insulators, are different type of topological state that emerges in certain materials with powerful spin-orbit coupling. Unlike the conventional QH effect, spin Quantum Hall states exhibit quantized Hall conductivity that is subjected to a spin degree of freedom rather than the charge. In spin-quantum Hall states, the Hall conductivity is quantized to the values that are half-integer multiples of conductance quantum, reflecting the presence of spin-polarized edge states. These states have different spins propagating in opposite directions along the edges, creating a spin-polarized current flow. Quantization of Hall-conductivity in spin quantum Hall states is associated with the spin Berry phase associated with the electron motion [14]. Spin quantum Hall states are a consequence of the interplay between powerful spin-orbit coupling or electron-electron interactions. The spin-orbit coupling induces spin splitting in the energy bands, and the electron-electron interactions further influence the formation of spin-polarized edge states.

2.2 Time-reversal symmetry (TRS)

Topological insulator (TI) exhibits time-reversal symmetry-protected curved metallic surface-states (SSs) that follow a Dirac-type linear energy and momentum dispersion relationship. Since the opposing spin channels remain locked to the opposing momentums during this operation, the topological insulator's helical surface states are unaffected. But this invariance and symmetry will be broken in the presence of MF and magnetic impurities. Symmetry-broken states produced in TI have been anticipated to convey numerous advanced quantum phenomena, including the Quantum Anomalous Hall effect, topological magneto-electric effect, and image magnetic monopole, despite the fact that topological insulator is an ordered phase that does not rely on broken symmetry. These exotic systems' special characteristics provide opportunities for both fundamental physics research and the development of novel materials with unusual properties for use in technology. By using ferromagnetic ordering, the instinctively broken time-reversal symmetry states might be experimentally inserted into a substance. Typically, two approaches can be used to accomplish this; doping via magnetic element and ferromagnetic proximity coupling [15].

Time reversal symmetry is a fundamental concept in topological insulators that plays a crucial role in their electronic properties. TSR refers to the invariance of a physical system under the reversal of the direction of time. In the context of topological insulators, it is a symmetry that protects the existence as well as stability of the surface states. The presence of the time-reversal symmetry in a material means that the laws of physics governing its behavior remain unchanged if the direction of time is reversed. Mathematically, this symmetry is represented by an operator called the time reversal operator (T). When applied to the Hamiltonian of a system, the time reversal operator reverses the momentum and the spin of the particles [16].

In the case of topological insulators, TRS is a crucial ingredient for the formation and protection of topologically protected surface states. These surface states, often known as topological surface states, are metallic states that exist within the energy gap of bulk insulating material. The presence of TRS ensures that the surface states in the topological insulator are robust against external perturbations and disorder. Any localized disorder or impurities that scatter electrons on the surface of the material must obey the TRS. This symmetry prevents backscattering, where electrons change their momentum and are scattered in the opposite direction, effectively protecting the surface states from being destroyed. The topological stability offered by TRS is closely associated with the unique features of surface states, such as their spin-momentum locking. In the time-reversal symmetric topological insulator, the spin orientation of surface states is tied to their momentum direction. This property ensures that backscattering processes that change the momentum of the electrons would also flip their spin, violating time-reversal symmetry. Moreover, the two counter-propagating modes are switched by time reversal, which also reverses the spin direction. The TRS is crucial in the understanding of topological surface states of insulators. The two counter-propagating modes are changed around via time reversal, which alters both the spin direction and the direction of propagation. Moreover, TRS plays a significant role in protecting the topological stability of such states. The edge of the "quantum spin Hall effect state" or two-dimensional TI contains right-moving and left-moving modes having an opposite spin and are closely related by the TRS. This edge can alternatively be thought of as one-half of the quantum wire, with electrons traveling in both directions in both spin-up as well as spin-down states [17]. Experimental techniques for example angle-resolved photo-emission spectroscopy (ARPES), can directly probe the TRS protected surface-states in topological insulators. These techniques provide detailed knowledge about the dispersion, momentum, and spin texture of the surface states, confirming their unique properties [18].

2.3   Topological surface-states

Topological surface-states, also known as topological edge states, are the distinctive feature of topological insulators. These surface states are electronic states which exist within the bulk band gap of an insulating material and exhibit unique properties, such as their robustness against disorder and their connection to nontrivial topological properties [19]. The key characteristic of topological surface states is their topological protection, which means they are insensitive to most external perturbations and imperfections. This protection arises from the combination of band topology and certain symmetries in the system, such as time reversal symmetry.

Properties

Robustness: Topological surface states are immune to scattering by impurities, defects, and other forms of disorder that would normally disrupt electronic states. This robustness arises from their topological nature, making them ideal for various applications in electronic devices and quantum information processing.

Spin-Momentum Locking: Topological surface states often exhibit a unique property referred to as spin-momentum locking. It means that the alignment of electron spin is intrinsically tied to its momentum direction. As a result, electrons moving in different directions on the surface of the topological insulator will have different spin orientations. This property has potential uses in quantum computing and spintronics [20].

Chiral Nature: Topological surface states can be chiral, meaning they propagate in only one direction along the material's edges or surfaces. This unidirectional propagation is an outcome of the nontrivial topology of the material's band structure and is robust against backscattering, enabling efficient and lossless transport of charge or spin.

Energy Dispersion: The energy dispersion of topological surface states can vary depending on the specific material and its band structure. In some cases, the dispersion can be linear, leading to a Dirac cone-like energy-momentum relationship resembling relativistic particles. In other cases, the dispersion can exhibit different shapes, such as quadratic or higher-order dispersions.

TIs are those quantum materials that exhibit a large band-gap-like normal insulator, but conducting surface-state which is topologically sheltered due to the combination of specific spin-orbit interaction and the TRS. Three-dimensional TIs surface states do have a striking resemblance to the 2D topological insulator's edge states. The direction of spin, which now fluctuates constantly as the function of the direction of propagation (fig. 4), nevertheless determines the direction of electron transport along the surface of three-dimensional TI, just as it did in the 2D case. The end result is an uncommon "planar metal" in which the direction of spin or propagation is fixed. Similar to the 2D situation, a three-dimensional topological insulator's surface states resemble half of a regular two-dimensional conductor and are topologically shielded from back-scattering. A three-dimensional topological insulator's surface permits electronic travel in any direction over the surface, while the direction of motion of the electron determines its direction of spin and vice versa. The "Dirac cone" structure of the 2D energy-momentum relation is evident [21].

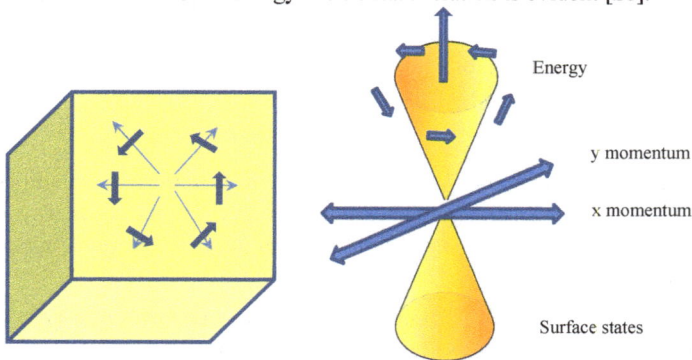

Figure 4. The surface of 3D topological insulator.

Experimental techniques for example angle-resolved photo-emission spectroscopy (ARPES) and scanning-tunneling microscopy (STM) have been instrumental in directly observing and characterizing the properties of topological surface states. These techniques provide insights into the dispersion, momentum, spin texture, and other properties of the surface states, confirming their topological nature. Topological surface states have collected significant interest due to their distinctive properties as well as potential uses in various fields, including electronics, quantum computing, and spintronics. They offer a fascinating platform for studying novel physical phenomena and exploring the fundamental properties of matter.

2.4   Spin orbital coupling

Spin-orbit coupling is a fundamental interaction that plays a significant role in the electronic properties of topological insulators. It arises from the coupling between the electron's spin and orbital motion, resulting in spin-orbit interaction that affects band structure and causes nontrivial topological states to emerge. In TIs, spin-orbit coupling is particularly significant because it contributes towards the formation of the unique topological surface-states (TSS). The spin-orbit coupling in these materials can be substantial due to the presence of heavy elements with powerful spin-orbit interactions, like bismuth and lead [22]. The spin-orbit coupling affects the electronic bands by lifting the spin degeneracy and creating spin-dependent energy shifts. Consequently, the electronic states near the band gap acquire a spin texture, where the direction of the electron's spin becomes tied to its momentum. This spin-momentum locking is a key aspect of topological surface states in TIs. The spin-momentum locking due to spin-orbit coupling has profound consequences. It makes the surface states robust against backscattering, as any scattering event that changes the momentum of an electron also flips its spin, violating the time-reversal symmetry. This protection against backscattering is a crucial aspect of the topological nature of the surface states and enables their propagation without significant energy loss. Additionally, the spin-orbit coupling in topological insulators can lead to interesting phenomena like the Rashba effect. The Rashba effect often refers to the spin splitting of electronic bands in the presence of structural inversion asymmetry. It can give rise to additional spin-textured states near the surface, enhancing the richness of the topological surface states [23]. Overall, spin-orbit coupling in TIs plays a dual role. On one hand, it affects the band structure, lifting the spin degeneracy and creating spin-textured surface states. On the other hand, it provides the protection mechanism that renders these states robust against backscattering, contributing to their topological nature. The interplay between spin-orbit coupling, band topology, and other symmetries gives rise to the unique properties and potential applications of topological insulators.

2.5   Bulk insulating states

Bulk insulating states in topological insulators refer to the insulating behavior exhibited by the bulk of these materials, while their surfaces or edges display metallic or conducting states. This bulk insulating property is a crucial aspect of topological insulators and is a result of the band structure and topological nature of these materials [24]. The bulk insulating states of TIs are characterized by a band-gap in bulk electronic band structure that separates the valence as well as conduction bands. This band gap is formed due to the distinctive electronic properties and topological nature of these materials. Here are some details regarding the bulk insulating states of topological insulators: Band Inversion and Topological Protection: The band gap in topological insulators is not a result of conventional electronic confinement but arises from a band inversion or band-crossing at specific points in the Brillouin zone. These band inversions occur between states with different symmetries or opposite parity, leading to the formation of a topological band gap. The presence of the band inversion is a crucial characteristic of topological insulators and is responsible for the protection of the topologically nontrivial surface or edge states. The band gap in the bulk separates these surface or edge states from the bulk states, ensuring their robustness against disorder and scattering.

Topological Invariants: The topological properties of TIs are often quantified by topological invariants, known as the Z_2 invariant. The Z_2 invariant is a topological quantum number that determines the absence or presence of protected surface states. The Z_2 invariant is associated with the parity of occupied electronic states in the bulk bands. A nontrivial Z_2 invariant indicates the presence of surface states, while a trivial Z_2 invariant corresponds to a fully gapped bulk insulator without any protected surface states.

Energy Band Structure: The bulk insulating states exhibit a distinct energy band structure. The conduction and valance bands are separated by a band gap, which can vary in magnitude depending on the specific material. The energy dispersion of the bulk bands can take different forms, such as Dirac cones, parabolic bands, or higher-order dispersions, depending on the material's symmetry and band structure. Hence, the energy band structure plays a fundamental role in determining the properties of topological surfaces or edge states, such as their dispersion, spin-momentum locking, and chirality.

Symmetry Considerations: The bulk insulating behavior of topological insulators is often associated with specific symmetries, such as TRS, inversion symmetry, or crystalline symmetries. These symmetries impose constraints on the electronic states, leading toward the formation of band gaps and the protection of the topological surface or edge states [25].

Experimental Characterization: The bulk insulating states and the associated band gap in topological insulators can be experimentally characterized using different techniques.

Angle-resolved photoemission spectroscopy (ARPES) provides specific information on the electronic band structure, energy dispersion, and Fermi surfaces. Scanning tunneling microscopy/spectroscopy (STM/STS) allows the direct visualization and probing of surface states and their local properties.

In summary, the bulk insulating states of topological insulators arise from band inversions, topological protection, and symmetry considerations. The band gap in the bulk separates the conduction and valance bands, while the topologically protected surface or edge states remain conducting. Understanding and characterizing the bulk insulating states is crucial for exploring the unique properties and potential applications of topological insulators.

2.6   Topological invariants

Topological invariants are mathematical quantities that capture the nontrivial topological properties of materials, including topological insulators. These invariants provide a rigorous framework for classifying different topological phases and determining the existence of topologically protected surface states or edge states. Here are some details regarding the topological invariants commonly used in the study of TIs:

Chern Number: The Chern number refers to the topological invariant used to characterize two-dimensional topological insulators, particularly those exhibiting the quantum Hall effect. It quantifies the topological properties of occupied energy bands in momentum space [26]. Mathematically, the Chern number is calculated as the integration of the Berry-curvature over the entire Brillouin zone. The Berry curvature measures the geometric phase acquired by an electron as it adiabatically moves through momentum space. A non-zero Chern number indicates the presence of topologically protected edge states and a quantized Hall conductivity.

Z_2 Invariant: The Z_2 invariant is a topological quantum number used to characterize three-dimensional topological insulators. It determines the absence or presence of topologically protected surface states. The Z_2 invariant takes values of 0 or 1 and is associated with the parity of occupied electronic states in the bulk bands. It can be calculated using different methods, such as the Fu-Kane-Mele method or the Wilson loop method. The Z_2 invariant of 0 corresponds to a trivial insulator without protected surface states, while a Z_2 invariant of 1 signifies a nontrivial topological insulator with robust surface states [27].

Fu-Kane-Mele (FKM) Invariant: The Fu-Kane-Mele invariant is a Z_2 invariant specifically designed for time-reversal invariant topological insulators. It characterizes the absence and presence of time-reversal symmetric surface states. The FKM invariant is calculated by the product of parities of occupied bands at time-reversal invariant momentum. A value of +1 indicates the presence of time-reversal symmetric surface states, while a value of -1 signifies the absence of such states.

Wilson Loop: The Wilson loop is a method used to calculate topological invariants in both 3D and 2D TIs. It involves tracing the evolution of a vector in the reciprocal lattice along a closed loop in momentum space. The accumulated phase of the vector along the loop gives information about the topological properties of a specific system. For example, in two dimensions, the Wilson loop can be used to calculate the Chern number, while in three dimensions, it can yield the Z_2 invariant.

They provide a robust framework for understanding the unique electronic properties and the existence of topologically protected surface states in these materials. The calculation and analysis of these invariants help in identifying and predicting the behavior of topological insulators and can guide new materials with specific topological properties.

3.   Fundamental properties of TIs

3.1   Photon-Like Electron

The dispersion relation in a typical conductor is non-linear whereas, a topological insulator is distinguished by a linear relationship between momentum and energy that behaves like a photon's propagation. Because of the high sensitivity of the topological surface towards the external applied electric field (EF), this characteristic can improve the performance of semiconductor devices. The generated average drift velocity in a unit EF represents the carrier conductivity and is known as carrier mobility. The ideal topological insulator has excellent mobility. The modulation doping technique has the potential to significantly improve mobility. The TI thus has enormous potential in this field. In actual use, a TI-based gadget with high mobility has a quick running speed as well as excellent cut-off frequency [28].

3.2   Low-Power Dissipation

In addition to the high mobility, the topological insulator has another outstanding property having low-power dissipation. The existence of different band gaps in an insulator causes resistance while, resistance in the metal is mainly due to electrons' collision with impurities, phonons, and many others. The surface of a topological insulator contains a Dirac electron that can evade the impurities and continue in its original path while coming in contact with it. Due to the electron's spin, the surface state of topological insulators has mainly four degrees of freedom which are twice as many as the one-dimensional system without having a spin. When it comes across the impurities, the electrons can move along the clockwise and counter clock-wise direction, and the direction of spin also changes. The principle of avoiding impurities is the coherent annulment of scattering waves. This may greatly decrease the resistance. In the meantime, an internal insulator prohibits the seepage of electricity. Hence, TI devices have the ability to operate at low power [29].

3.3   Spin-Polarized Electrons

The surface state of TI has spin-polarized electrons. The spin has two directions—up and down—and this adds to the degree of flexibility. As a result, the spin-polarized electrons can be directed in various ways. ARPES can distinguish the spin-polarized Dirac-cone electrons on the TI surface state. The surface of the topological insulators has spin-dependent electrons with spin-momentum locking despite having an insulating bulk state. As a result, this also enhances the understanding of magnetic and spin-electronic devices.

3.4   Quantum Spin Hall (QSH)

The most striking TI asset is the QSH. The half-integer QH effect can be studied using the surface state of TIs, which can also be used to conduct QSH experiments. Additionally, it is possible to study magnetic mono-polarity and fractional charge. Due to the instability of the regular quantum state, de-coherence will occur while monitoring it. After modest disturbances, the amplitude of wave-function probability instantly switches from an ordered to a discrete distribution. However, the TI's quantum state is extremely stable and unaffected by even little disturbances, making it possible to use TI for quantum processing [30].

3.5   Mechanical strength

Many topological insulators, especially those based on heavy elements such as bismuth and antimony, are brittle in nature. Brittle materials have a tendency to fracture or break under stress without significant plastic deformation. This property can limit their use in certain applications that require materials with high mechanical toughness or flexibility. These materials typically possess a crystalline lattice structure, and the arrangement of atoms within the lattice can influence their mechanical behavior. The crystal structure may determine factors such as hardness, elastic modulus, and fracture toughness.

3.6   Thermal Expansion and Mechanical Stability

The coefficient of thermal expansion is the measurement of how much a material can expand or contract by changing temperature. The thermal expansion behavior of topological insulators can be influenced by their crystal structure and chemical composition. Understanding the thermal expansion properties is important for applications that involve temperature variations, as it can affect the stability and reliability of devices.

Topological Insulators: Materials and Applications   Materials Research Forum LLC
Materials Research Foundations 154 (2024) 1-20   https://doi.org/10.21741/9781644902851-1

The mechanical stability of a material refers to its ability to maintain its structure and integrity under external mechanical loads. While the electronic properties of topological insulators are often the primary focus, their mechanical stability is also important for practical applications. It ensures that the material can withstand mechanical stresses without undergoing irreversible structural changes or failure [31].

3.7 Band inversion and Dirac-like surface-states

The electronic band structure of TIs undergoes a band inversion compared to ordinary insulators. In the bulk, the energy bands are arranged such that the conduction band and valance band can overlap at certain momentum points in the Brillouin zone. This band inversion is crucial for the emergence of the conducting surface or edge states in TI. Moreover, the conducting surface as well as edge states in TIs exhibits a Dirac-like dispersion relation. This means that the energy-momentum relationship for these states is linear, resembling the behavior of mass-less Dirac fermions. The Dirac-like surface states contribute to the distinctive electronic properties and unusual transport phenomena observed in topological insulators [32].

4. Development of TIs

Topological insulators (TIs) are those materials that have insulating properties in their bulk but possess conducting states on their surfaces or edges. These surface or edge-states are topologically protected, meaning they are robust against perturbations and disorder. The field of topological insulators gained significant attention with the experimental discovery of two-dimensional (2D) topological insulators in 2007 and three-dimensional (3D) TIs in 2008. The pioneering works by groups led by Konstantin Novoselov and Shoucheng Zhang, respectively, opened up a new research direction.

The 2D topological state was the preliminary stage of the development of TIs. The uniform incline of the 2D semiconductor system and two-dimensional substances like graphene possibly led to this state. Making the material have bulk band inversion is the fundamental method for obtaining TI. This idea was effectively purposed by B.A. Bernevig, T. Hughes, and S.C. Zhang in 2007. In a CdTe/HgTe/CdTe semiconductor quantum well, the HgTe layer's thickness can be altered to achieve the QSH. This was performed by the CdTe's spin-orbit coupling effect, which was not very strong. The HgCdTe system's spin-orbit coupling was equivalent to being improved by thickening the HgTe layer. The p-orbital was pushed over to the s-orbital in the system when thickness reached roughly 6.5 nm due to the bulk band-inversion [33].

The third generation of topological insulators is known as topological-crystalline insulators (TCIs). TCIs were synthesized through the calculation of the structure of the band gap as well as topological-band analysis in 2013. There was mainly a boundary state having spin-filter characteristics sheltered through mirror symmetry onto the edges. Instead of time-reversal symmetry, they possessed an even numeral of Dirac cones that were secured by the lattice's mirror symmetry. The films of SnTe and $Pb_xSn_{1-x}Se(Te)$ allowed for the achievement of this unique topological phase. An even better feature of TI's third generation was the ability to regulate the band gap. The mirror symmetry of the system was broken, leading to a controlled band gap on the edge state, by manipulating the EF that is perpendicular to the film. Additionally, the bulk has a tiny band gap while the surface has no gaps. TCIs are a type of promising material since they might be employed in numerous devices, including photo-detectors. In addition to 2D and 3D TIs, the concept of higher-dimensional topological insulators has emerged. These materials possess conducting surface states that are confined by higher-order topological invariants. For example, higher-dimensional analogs of the Quantum Spin Hall effect, such as the higher-order topological insulators and higher-order Dirac semimetals, have been projected and studied. It is important to note that the field of TIs is highly active and rapidly evolving. Researchers continue to investigate new materials, explore novel phenomena, and seek ways to harness the unique properties of topological insulators for technological advancements.

Conclusion

This chapter summarizes the fundamental concepts, terminologies, and development of TIs. These materials, which are insulating in their bulk but hold edge-states and conducting surfaces protected by topology, have unique electronic properties that hold great promise for both fundamental research and technological applications. Researchers have identified different materials, such as bismuth-antimony, bismuth-selenide, and bismuth-telluride, as topological insulators. The field of TIs is a dynamic and prominent area of research. The development of topological insulators suggests the possibilities of transforming electronics, spintronics, and quantum computing.

References

[1] X. Zhang, J. Wang, S.C. Zhang, Topological insulators for high-performance terahertz to infrared applications, Phys. Rev. B 82 (2010) 245107. https://doi.org/10.1103/PhysRevB.82.245107

[2] C.O. Taberner, M. Hermanns, From Hermitian critical to non-Hermitian point-gapped phases, arXiv Prepr. arXiv2211 (2022) 13721.

[3] L.H. López, Searching for kagome multi-bands and edge states in a predicted organic topological insulator, Nanoscale 13 (2021) 5216-5223. https://doi.org/10.1039/D0NR08558H

[4] M. He, H. Sun, Q.L. He, Topological insulator: Spintronics and quantum computations, Front. Phys. 14 (2019) 1-16. https://doi.org/10.1007/s11467-019-0893-4

[5] A. Lawal, Theoretical study of structural, electronic and optical properties of bismuth-selenide, bismuth-telluride and antimony-telluride/graphene heterostructure for broadband photodetector, Universiti Teknologi, Malaysia, 2017.

[6] K.V. Klitzing, The Quantum Hall Effect, Springer Science & Business Media, 2012.

[7] M.Z. Hasan, C.L. Kane, Colloquium: Topological insulators, Rev. Mod. Phys. 82 (2010) 3045. https://doi.org/10.1103/RevModPhys.82.3045

[8] A.K. Cheetham, R. Seshadri, F. Wudl, Chemical synthesis and materials discovery, Nat. Synth. 1 (2022) 514-520. https://doi.org/10.1038/s44160-022-00096-3

[9] J. Estes, A. O'Bannon, E. Tsatis, T. Wrase, Holographic Wilson loops, dielectric interfaces, and topological insulators, Phys. Rev. D 87 (2013) 106005. https://doi.org/10.1103/PhysRevD.87.106005

[10] X.L. Qi, S.C. Zhang, The quantum spin hall effect and topological insulators, arXiv Prepr. arXiv1001 (2010) 1602.

[11] C. Cao, J. Chen, Quantum spin hall materials, Adv. Quantum Technol. 2 (2019) 1900026. https://doi.org/10.1002/qute.201900026

[12] K.V. Klitzing, The finite-structure constant α: A contribution of semiconductor physics to the determination of α, Festkörperprobleme 21 (1981) 1-23. https://doi.org/10.1007/BFb0108600

[13] K.V. Klitzing, 25 Years of Quantum Hall effect (QHE): A personal view on the discovery, physics, and applications of this quantum effect, in: The Quantum Hall Effect: Poincaré Seminar, 2004, pp. 1-21. https://doi.org/10.1007/3-7643-7393-8_1

[14] B. Basu, P. Bandyopadhyay, Edge states of Quantum Hall fluid and Berry phase, Int. J. Mod. Phys. B 11 (1997) 2707-2726. https://doi.org/10.1142/S0217979297001337

[15] C.Z. Chang, P. Wei, J.S. Moodera, Breaking time-reversal symmetry in topological insulators, MRS Bull. 39 (2014) 867-872. https://doi.org/10.1557/mrs.2014.195

[16] M. Sigrist, Time-reversal symmetry breaking states in high-temperature superconductors, Prog. Theor. Phys. 99 (1998) 899-929. https://doi.org/10.1143/PTP.99.899

[17] M.Z. Hasan, S. Xu, M. Neupane, Topological insulators, topological Dirac semimetals, topological crystalline insulators, and topological Kondo insulators, Wiley, 2015, pp. 55-100. https://doi.org/10.1002/9783527681594.ch4

[18] X. Zhou, C. Fang, W.F. Tsai, J. Hu, Theory of quasiparticle scattering in a two-dimensional system of helical Dirac fermions: Surface band structure of a three-dimensional topological insulator, Phys. Rev. B 80 (2009) 245317. https://doi.org/10.1103/PhysRevB.80.245317

[19] L. Müchler, Topological insulators from a chemist's perspective, Angew. Chemie Int. Ed. 51 (2012) 7221-7225. https://doi.org/10.1002/anie.201202480

[20] P. Corbae, Observation of spin-momentum locked surface states in amorphous Bi2Se3, Nat. Mater. 22 (2023) 200-206. https://doi.org/10.1038/s41563-022-01458-0

[21] C. Kane, J. Moore, Topological insulators, Phys. World 24 (2011) 32. https://doi.org/10.1088/2058-7058/24/02/36

[22] E.E. Krasovskii, Spin-orbit coupling at surfaces and 2D materials, J. Phys. Condens. Matter 27 (2015) 493001. https://doi.org/10.1088/0953-8984/27/49/493001

[23] A. Manchon, H.C. Koo, J. Nitta, S.M. Frolov, R.A. Duine, New perspectives for Rashba spin-orbit coupling, Nat. Mater. 14 (2015) 871-882. https://doi.org/10.1038/nmat4360

[24] J.E. Moore, The birth of topological insulators, Nature 464 (2010) 194-198. https://doi.org/10.1038/nature08916

[25] J. Kruthoff, J. de Boer, J. van Wezel, Topology in time-reversal symmetric crystals, Phys. Rev. B 100 (2019) 75116. https://doi.org/10.1103/PhysRevB.100.075116

[26] J. Ge, High-Chern-number and high-temperature quantum Hall effect without Landau levels, Natl. Sci. Rev. 7 (2020) 1280-1287. https://doi.org/10.1093/nsr/nwaa089

[27] J.C.Y. Teo, L. Fu, C.L. Kane, Surface states and topological invariants in three-dimensional topological insulators: Application to Bi(1− x) Sbx, Phys. Rev. B 78 (2008) 45426. https://doi.org/10.1103/PhysRevB.78.045426

[28] S.K. Mishra, S. Satpathy, O. Jepsen, Electronic structure and thermoelectric properties of bismuth telluride and bismuth selenide, J. Phys. Condens. Matter 9 (1997) 461. https://doi.org/10.1088/0953-8984/9/2/014

[29] W. Tian, W. Yu, J. Shi, Y. Wang, The property, preparation, and application of topological insulators: A review, Materials (Basel) 10 (2017) 814. https://doi.org/10.3390/ma10070814

[30] X.L. Qi, R. Li, J. Zang, S.C. Zhang, Inducing a magnetic monopole with topological surface states, Science 323 (2009) 1184-1187. https://doi.org/10.1126/science.1167747

[31] R. J. Cava, H. Ji, M. K. Fuccillo, Q.D. Gibson, Y.S. Hor, Crystal structure and chemistry of topological insulators, J. Mater. Chem. C 1 (2013) 3176-3189. https://doi.org/10.1039/c3tc30186a

[32] Z. Yue, X. Wang, M. Gu, Topological insulator materials for advanced optoelectronic devices, arXiv:1802.07841 (2019) 45-70. https://doi.org/10.1002/9781119407317.ch2

[33] B. Yan, S.C. Zhang, Topological materials, Reports Prog. Phys. 75 (2012) 96501. https://doi.org/10.1088/0034-4885/75/9/096501

Topological Insulators: Materials and Applications
Materials Research Foundations 154 (2024) 21-46

Materials Research Forum LLC
https://doi.org/10.21741/9781644902851-2

Chapter 2

One-Dimensional Topological Insulators

M. Rizwan[1*], T. Hashmi[1], A. Ayub[2]

[1]School of Physical Sciences, University of the Punjab, Lahore, Pakistan

[2]Department of Physics, University of Gujrat, Gujrat, Pakistan

*rizwan.sps@pu.edu.pk

Abstract

One-dimensional topological insulators have garnered quite a lot of attention in recent times. These topological insulators (TIs) are crucial in the comprehension of topological properties. This chapter provides a very detailed and comprehensive overview of these astonishing materials. From history, classification based on symmetry, Dirac points, and dimensions, generations such as 1st, 2nd, and higher order TIs, synthesis techniques such as physical vapor deposition, chemical vapor deposition, 2D, 3D Tis and future models of these topological insulators, all are deliberated in detail. Future aspects are discussed as well, this chapter is composed to fully enlighten the reader on these 1D topological insulators.

Keywords

Spin-Polarized Electrons, 1D Topological Insulators, Physical Vapor Deposition, Time-Reversal Symmetry, Photonic Topological Insulators

Contents

Topological Insulators: Materials and Applications   Materials Research Forum LLC
Materials Research Foundations 154 (2024) 21-46   https://doi.org/10.21741/9781644902851-2

1.   Introduction

Dealing with the microscopic world has become easier with the advancement of science and technology, especially the physics of condensed matter and nanoparticles. Phases along with their distribution are a key field in this perspective. Before the invention of topological insulators, phases of matter were mostly concerned with breaking symmetries. Consequently, topological insulators do not fall into this category. For many semiconductor devices, this innovation of topological insulators is very fruitful. After the discovery of graphene, topological insulators are another important discovery that will certainly drive the expansion of science [1], 1D topological insulators are a particular genre of materials that exhibit special electrical behavior due to their topological nature. These materials often have a chain-like or thread-like pattern that makes them look the same. They differ from traditional insulators with topologically protected edge states that prevent impurities and problems. Because of the interactions between strong spin-orbit coupling and T-symmetries, edge states located on the substance's surface occur in 1D topological insulators. These are responsible for materials' unusual behavior as compared to aggregated states. At different energy states, gapless edge stat in 1D topological insulators, which is a unique property of this material. Edge states are usually undistributed or have linear distribution and are protected by topology devices. This means that these edges are very sensitive to local perturbations or impurities making the highly desirable for implementations in quantum devices and electronics. The existence of these barriers leads to interesting phenomena like self-transmission and nonlocal transport making 1D topological insulators promising candidates for many applications in quantum computing, spintronics, and other technologies. The basic idea behind the discovery of topological insulators is the Quantum Hall effect (QH), which is the Hall effect at low temperatures under a strong magnetic field. Although QH is a great discovery its development is limited as it requires a high magnetic field and low-temperature conditions which involves high cost. This problem was resolved in 2005 by the discovery of the Quantum spin hall effect (QHS) as it exhibits the same electronic states as QH as a result of strong spin-orbit coupling. The finding opens new doors for scientific research [2].

1.1   Overview of TIs

Solid Matter is usually split up into three broad groups: Conductors, insulators as well as semiconductors. A conductor is a material or a medium that allows the flow of electric charge or heat. In the context of electricity, a conductor is a material that allows electric charge to flow through it easily. Examples of conductors include metals, such as copper and aluminum, as well as some liquids and solutions that contain ions. Conductors are essential components in electrical circuits, as they are used to transmit electricity from one point to another. Free electrons move without interference. Insulators are materials that do not allow the flow of electric charge or heat easily. Unlike conductors, which have low electrical resistance or high thermal conductivity, insulators have high electrical resistance and low thermal conductivity. This means that insulators restrict the flow of electric charge or heat, effectively blocking or inhibiting their transfer through the material. Insulators are commonly used in electrical and thermal insulation applications where it is desirable to prevent the flow of electric charge or heat. The properties of insulators arise from their atomic or molecular structure. Insulators typically have tightly bound electrons in their atomic or molecular structure, which do not move easily in response to an electric field or temperature gradient. This lack of free or mobile charge carriers makes insulators poor conductors of electricity or heat. The intermediate state between conductor and insulators is a semiconductor.

Consequently, topological insulators do not fall among any of these materials. It has a huge electronic state with confined bandwidth indicating the absence of free carriers and has conductive metallic surfaces that have Dirac points that can easily pass through the bandwidth [3]. The material that acts as an insulator and its surface act as an electrical conductor is called a topological insulator. This means that the movement of electricity can only be seen on the surface. 1D topological insulators (1D TIs) are materials that exhibit special electrical properties due to their topological properties, unlike ordinary insulators or conductors. These materials differ from each other in electron conduction but have topologically protected active endstates, that is, they are resistant to collision and vice versa.

These materials have a band gap in the bulk, preventing electron conduction, but have a robust conducting edge state that is topologically protected meaning they are resistant to disorder and backscattering. As in ordinary or identical electrical materials, in topological insulators, there is a deviation between the energies of valence and conduction bands, but in ordinary insulators, these bands are broken or apart [4]. Without untwisting the bands, one cannot transform the topological insulator into a standard insulator. Since the bands are not bent, closing the band gap, allows conduction to occur. Due to its uniqueness, local interference does not affect the surface state of the topology band structure. Although ordinary insulators can also support surface activity, only Tis have this flexibility [5].

2. History

The concept of topological insulators was first proposed by theoretical physicists Duncan Haldane and Charles Kane in the early 2000s, and the field has grown rapidly since then. Below is a brief history of key points in the discovery and development of topological insulators: In 1985 Volkov and Pankratov proposed the first 3-dimensional model of topological insulators.

It was later proposed by Pankratov, Pakhomov, and Volkov in 1987 [6]. In 2005 Duncan Haldane proposed the concept of quantum spin hall (QHS) insulators, which are 2D topological insulators with a time-defying spin-polarized edge state. This concept laid the foundation for the study of topological insulators and spurred further research and experimentation. The existence of three-dimensional topological insulators, also known as "3D strong topological insulators" or Z2 invariants were presented by Kane and Eugene. These materials have been shown to have solid states, have a Dirac-like distribution, are protected by inverse time symmetry, and exhibit unique properties. In 2007, a research team led by Professor Zhang Shoucheng of standard university confirmed the theoretical prediction of QHS insulators. We thank Charles Kane and Professor Zhang for their theoretical advice and experimental work on topological insulators. This information demonstrates the significance and impact of topological insulators in the field of condensed matter. Since their initial discovery, topological insulators have attracted researchers worldwide, which has led to experiments and research. These materials show great potential in many applications, including quantum computing, spintronics and topological quantum phenomena. These topological materials play an important role in physics research and many semiconductor devices. After the discovery of graphene, topological insulators are an important discovery that will certainly increase the power of scientific research. Both the dimensions of the material and its fundamental symmetry tell us about the properties and surface states of topological insulators. At room temperature, topological insulators have a QSH state (created in 2005), which has led to their use and engineering [7].

3. Properties

These materials have shown great promise for various applications, including quantum computing, spintronics, and topological quantum phenomena. Since the initial discovery, topological insulators have attracted significant attention from researchers worldwide, leading to a flurry of experimental and theoretical investigations. Ongoing research continues to uncover new properties and potential applications of topological insulators, making them a fascinating and rapidly evolving field of study in condensed matter physics. The material's phase is generally expressed by the impulsive splitting of symmetry before the discovery of topological insulators. However, symmetric breaking theory, local order parameters, and long-range correlation should not describe the topological order of topological insulators as they have no concern with them. Topological insulators play an important role in world exploration. A topological insulator is a notable innovation off the back of graphene and it will definitely escort to the enhancement of investigation in the scientific world. Both dimensions of the material and its elemental symmetries tell us about the premises of topological insulators along with their surface states [8]

3.1 Photon-like electron

Compared to the nonlinear distribution relationship of human conductors, topological insulators are characterized by a linear distribution relationship between power and energy, similar to the propagation of photons. Due to the high sensitivity of the topological surface to external electricity. The performance of semiconductor devices is improved by this device [9]. Carrier conductivity is represented by the average drift velocity (carrier mobility) per unit of electric power. Supreme topological insulators are highly mobilized. Theoretically, the Order of magnitude of the mobility is increased by the modulation doping technique. In practical application, high mobility leads to high operating speed and greater cut-off frequency of topological insulator-based devices.

3.2 Low power dissipation

Topological insulators have another amazing property i.e., low power dissipation. The opposition of insulators is because of the band gap and the bumping of electrons with photons and other impurities causes the metal resistance. But the surface electrons i.e., Dirac electrons of the topological insulator move in the first direction and can pass the impurity when encountering the impurity. Because the electrons are two times greater as a spin-less one-dimensional network, the surface state has four degrees of freedom. When an electron encounters an impurity, it moves clockwise or counterclockwise and its spin can be reversed. The overall effect of impurities is thus avoided, resulting in much less resistance and no leakage from the inner insulator. Therefore, materials based on topological insulators can operate at less power.

3.3 Spin-polarized electrons

As discussed earlier, the topological insulators have four degrees of freedom. This number can be increased by the spin polarization of Dirac electrons. In order to bind the electron in specific directions only, the spin of an electron is split into two directions (up and down). The spin of the Dirac cones can be observed with the help of Angle-resolved photoelectron spectroscopy (ARPES). Although the topological insulators have an insulating state, the exterior has spin-reliant electrons which are orthogonal to momentum which leads to the information of spintronic.

3.4   Quantum spin hall effect (QSH)

QSH is the most important property of topological insulators. With QSH, experiments can be done within the exterior state of topological insulators, along with the quasi-quantum hall effect, magnetic monopoles, and low charges [10]. Due to the instability of normal quantum states, inconsistencies may occur when observing the quantum states. Immediately after a slight perturbation, the wave function amplitude changes from a continuous to a discrete distribution. However, these minor problems do not affect the quantum state of the topological insulators as it is very stable and enables quantum computing. These materials are 2D topological insulators with the spin-momentum interlocked conductive states, i.e., the spin orientation of the electron is locked into their momentum orientation. This unique property makes quantum spin hall insulators attractive for spintronics applications. HgTe/CdTe quantum well is the first theoretically discovered quantum spin Hall insulator and was experimentally verified in 2007.

4.   Class distribution of TIs

Topological insulators are divided into three according to size, symmetry of Dirac points, determining whether the topological insulator is strong or not, and their equality.

4.1   Distribution by dimension

Topological insulators are classified into 1D, 2D, and 3D structures. The two-dimensional structure has a layer with large voids, and the ideal quantum structure is two-dimensional. These materials are generally insulators and have protected states. The edges of the 2D topological insulators are now 1D along the edge of the backscatter-resistant material. The first 2D topological insulator discovered is the HgTe quantum well, which was theoretically and experimentally verified in 2007. 3D topological insulators are more promising than 2D topological insulators due to their stable stoichiometry and pure chemical phases. The present compounds are easily produced and have a moderate band gap (0.35ev) [11]. The quantum and physical properties of Weyl semimetals (WSMS) allow the study of chiral anomalies, Fermi arcs, and other phenomena in condensed matter physics. 2D topological insulators have a single electronic state, and 3D topological insulators have a two- dimensional electronic state, i.e., the outside is metal, and the body is an ordinary insulator. The reason for the difference between the exterior state and the interior is that the edge state follows a transition path between the exterior and bulk.

These materials are very insulating, i.e., they do not conduct electricity from the inside but have a surface layer that conducts electricity. The external state of 3D topological insulators is topologically conserved, meaning they protect the mysterious and unshattered electrons. The first 3D topological insulator found is a bismuth antimony metal (BiSb), which was experimentally proven in 2007. Currently, one dimensional (1D) topological level is attracting more attention. Topological properties can be easily discussed using one-dimensional and three-dimensional systems. The 1-dimensional system exhibits a rich topological equation due to its high symmetry. This method has a simple structure, is not very complicated, and can be used experimentally. For this reason, additional topological features can be easily checked. Some topological properties of 1-Dimensional topological insulators have been studied so far using cold atoms or photonic crystal setups. Nowadays one-dimensional topological phases are getting more attention. Topological properties can be easily discussed by using one-dimensional systems, like two and three-dimensional ones.

4.2   Distribution by parity of Dirac points

Topological insulators can be split into two groups according to Dirac point's parity: strong or powerful topological insulators and weak or fragile topological insulators. If the Brillouin surface has an even digit of Dirac points, then it is said to be a weak topological insulator. Strong collisions ensure good placement of electronic material on the surface [12]. For strong topological insulators, the Brillouin domain has an odd digit of Dirac points. The perfect metal state is achieved because the disturbance does not affect the position of the electrons on the surface. This also explains why only a small fraction of compounds with spin-orbit coupling are topological insulators. In general, Topological insulators are strong because their molecular structure is almost similar.

4.3   Distribution by symmetry

Topological insulators (TIs) are typically distributed with respect to their symmetries, which are the intrinsic geometric and mathematical properties that direct their behavior. Due to their time-reverse symmetry and low crystallographic symmetry, WSMs collectively have Weyl fermions and Fermi arcs at the center. The Weyl point of type 2 WSM is formed at the boundary between electron and pocket as it does not follow Lorentz symmetry compared to type 1 WSM. There are various ways to sort TIS based on symmetries here are some common approaches.

- **Time-reversal symmetry (TRS):** TIs can be classified into TRS-protected or TRS-breaking TIs. TRS is a fundamental symmetry that dictates how a system behaves under time reversal, which involves reversing the direction of time. TRS-protected TIs have an odd number of Dirac cones or helical edge states protected by TRS, which ensures their stability against backscattering. TRS-breaking TIs, on the other hand, do not have this protection and may exhibit different types of surface states.

- **Inversion symmetry (IS):** TIs can also be classified into inversion-protected or inversion-breaking TIs. Inversion symmetry refers to the symmetry of a system under inversion, which involves swapping the coordinates of a point with its negatives. At certain high-symmetry points in the Brillouin zone, inversion-protected TIs have inverted band structures resulting in gapless surface states. Inversion-breaking TIs lack this symmetry and may have different surface properties.

- **Mirror symmetry (MS):** TIs can be classified based on mirror symmetries, which refer to the reflection symmetry of a system along certain planes. Mirror-protected TIs exhibit protected surface states that are localized along the mirror planes while mirror-breaking TIs lack this symmetry and may have different surface states.

- **Rotational symmetry:** TIs can also be classified based on their rotational symmetries, which refer to the rotational symmetry of a system around certain axes. Rotational symmetries can affect the nature of surface states and their dispersion.

5. Synthesis of TIs

Topological insulators can be synthesized by multiple methods. Each method has advantages and disadvantages. In practice, low-cost preparation methods can produce high purity topological insulators. The properties of topological insulators also depend on doping because impurities cannot be avoided during preparation. These impurities can be manipulated to obtain certain properties. For example, n-type and p-type three-dimensional topological insulators can be obtained by controlling the amount and charge of the doping material. Topological insulators are produced by mechanical exfoliation, molecular beam epitaxy (MBE), physical vapor deposition (PVD), chemical vapor deposition (CVD), Solvothermal synthesis, and many more. Weak van der Walls interactions are responsible for the development of thin-film topological insulators. These fragile interaction causes a thin film of crystal to peel off with clean and perfect surfaces [13]. Van der Waals epitaxy (VDWE) is a good way to grow topological insulators layer-by-layer over other substrates for heterostructures-integrated structures.

5.1 Mechanical exfoliation

The 3D topological insulator has a unique molecular structure. Its simple QL consists of two Bi atoms and three Se atoms bonded side by side with weak van der Waals forces. Under the influence of external force, the crystal is easily separated from layer to layer.

Graphene is obtained by exfoliating graphite layer by layer. Similarly, the exfoliation of topological insulators produces a thin film of topological insulators. Hong et al.

successfully prepared ultra-thin Bi_2Se_3 thin films (nanobelts) in 2010, using the vapor-liquid-solid (VLS) method.

Ultrathin Bi_2Se_3 nanoribbons of very low width (QL at least 1 QL) were exfoliated from thick Nano ribbons (over 50 QL) with the help of an atomic force microscope (AFM). It is a fully controlled mechanical exfoliation method [14].

5.2   MBE growth of TIs

Semiconductor and photovoltaic thin films are produced by molecular beam epitaxy. An ordered Layer of crystalline material on a crystalline substrate is synthesized by using an epitaxy method called MBE. A high vacuum or ultra-high vacuum is required for MBE. One or more elements of the thermal atomic (heated until sublimation in a different electron beam evaporator) or molecular beam are injected onto the heated substrate surface in an ultra-high vacuum. These molecular beams interact with the surface of the substrate and form a single crystal thin film. MBE is the most effective method for developing single crystals where molecular beams lead to surface migration, adsorption, and nucleation after energy exchange with the substrate. Horizontally transforms into movies. This reaction involves both chemical and physical changes. During formation, the compound binds to the substrate. The MBE is made in a high vacuum with minimal contamination. However, a large lattice mismatch and defects are also observed at the interface, which can be overcome by controlling the growth rate [15]. Any change in growth rate will also lead to different types of carriers, so the growth rate must be improved again to improve the mobility of the material. These materials have flat surfaces and smooth interfaces because MBE samples can be grown in layers and can easily transfer from the growing chamber to the characterization, chamber like ARPES and STM studies. Topological insulators can be developed on a broad range of substrates due to weak Van der Waals bonding such as Silicon Si, Aluminum Oxide Al_2O_3, Gallium arsenide GaAs, and many more.

5.3   Chemical vapor deposition

Chemical Vapor Deposition (CVD) is a technique for fabricating topological insulators, a kind of matter that have special electrical properties on their surfaces. Topological insulators are materials having insulating properties as well as a conductive surface that prevents topological cracking. The CVD method [16] for preparing topological insulators generally includes the following steps:

Selection of starting material: These materials are selected according to the desired topological properties. These precursor materials are usually organic or inorganic compounds that can easily evaporate and be transferred to the substrate surface.

Preparation of Substrate: The substrate, usually a crystal or a thin film of suitable material, is cleaned and prepared to provide a suitable surface for the release of the primary material. This may include polishing, etching, or other surface preparation to provide a clean and smooth substrate surface.

Loading of the pre-material: The pre-material is transported to a suitable location, such as a generator or gas stream, which allows controlled evaporation and transport of the pre-material to the substrate.

Deposition process: The substrate is placed in the reaction chamber and then the reaction chamber is discharged into a low-pressure environment to create a controlled atmosphere for process deposition. The starting materials are heated to evaporate and transfer to the substratum surface. The starting material then reacts with the substratum surface to form a topological insulator material.

Control of process parameters: Various process parameters such as temperature, pressure, and flow rate are carefully controlled to obtain the desired composition, thickness, and crystal structure of the topological insulator material. This will require optimization by trial and error to achieve the desired results.

Postdeposition treatment: After deposition, topological insulating materials can be treated such as annealing or deposition functionalization after deposition to improve their properties or remove residual impurities.

Characterization: Finally, the prepared topological material is characterized using different expertise such as XRD, SEM, and TEM, and its presence, composition, structure, and electrical properties are determined. In general, the CVD method for the preparation of topological insulators, spintronics, quantum computing, etc. has a broad and controlled approach to producing interesting materials with properties suitable for applications in advanced electronics.

5.4 Physical vapor deposition (PVD)

Physical Vapor Deposition (PVD) is another technique used to create topological insulating materials. The PVD method involves depositing the material onto the substrate by physical means such as spraying or evaporation. Without any chemical reaction. Below are the general steps of the PVD method for the production of topological insulators. The substrate (usually a crystal or a thin film of suitable material) is cleaned and prepared to a clean and smooth surface for precipitation. Then load the substrate into the PVD deposition chamber. The topological insulating material is usually prepared as a target and transported to the appropriate location in the PVD chamber. The base is then heated using methods such as resistive heating or electron beam bombardment to create a vapor from the topological insulating material [17]. The steam produced is taken to the substrate surface, where it condenses and forms a thin film of topological insulating material. The deposition can be done by various PVD techniques such as evaporation or sputtering depending on the specific material and thin film properties. Various process parameters such as temperature, pressure, and discharge rate are carefully controlled to obtain the desired composition, thickness, and crystal structure of the topological insulating film. These parameters must be optimized to achieve the desired properties. Film growth during deposition can be monitored in situ using techniques such as RHEED or Quartz Crystal Microbalance QCM, which provides real feedback to control deposition and refine the film. After deposition, the topological insulator thin film can be processed by annealing or surface functionalization to improve its properties or remove residual properties. The released topological insulating films are then used to determine their composition, structure, and properties using electricity and various techniques. The PVD method has the flexibility to control negative growth and can produce a good topological insulating film with good properties. They are widely used in research and manufacturing to create topological insulating materials and to explore their unique properties for advanced electronics applications.

6.   Generations of TIs

Topological insulators belong to a special group of materials that offer special electrical properties due to their non-trivial determination. Over the years, researchers have identified several generations or classes of topological insulators based on their physical and experimental properties.

6.1   First-generation TIs

First-generation topological insulators are predicted theoretically and discovered experimentally in two-dimensional (2D) systems. This material is completely insulating but has a conductive surface state i.e., preserved by time reversal anomalies. The first proven system of first-generation topological insulators is the HgTe/CdTe quantum well system, which was tested to demonstrate the QHS effect in which opposite spin counter-propagating edge states are topologically conserved [18]. First-generation topological insulators refer to a class of materials that began to be seen and studied in the late 2000s and early 2010s. These materials are characterized by their unique electronic properties and are protected by time-reverse symmetry, and exhibit topologically conserved surface states. The discovery of topological insulators along with applications in fields such as quantum computing and spintronics opened up a new field in condensed matter physics. A new generation of topological insulators has also been discovered, including high-order topological insulators and topological crystal insulators, which exhibit additional properties and behaviors. The field of topological insulators is a rapidly growing research field with exciting potential for new discoveries and technological applications [19].

6.2   Second-generation TIs

Second-generation topological insulators contain three-dimensional (3D) materials with topologically conserved surface states. This knowledge is characterized by the connection of the circle with the strong, resulting in the difference between the difference and the strong. Bi_2Se_3, Bi_2Te_3, and Sb_2Te_3 are some examples of second-generation topological insulators which have been extensively studied due to their wide band gaps and topologically conserved surface states, making them attractive for many applications in spintronics, quantum computing, and other quantum technologies. Second-generation topological insulators (TIs) are part of a group of materials that have unique electronic properties and are part of the general field of condensed matter physics. TIs are materials that act as complete insulators but have topologically protected surface layers, i.e., resistant to impurities and impurities. First-generation TIs were discovered in the early 2000s and have linear distribution which leads to special properties of TIs such as the QSH effect and having a helical edge province. However, these materials are often limited by poor performance, lack of constraints, and problems in producing good-quality materials [20].

Second-generation TIs are a newer class of TIs that have emerged in recent years and are designed to overcome some of the limitations of first-generation TIs. These materials typically exhibit more complex surface states, such as multiple Dirac cones or higher-order topology, which can lead to additional exotic electronic properties. They are also engineered to have improved tunability, higher operational temperatures, and enhanced device performance. Second-generation TIs are being actively researched due to their potential for various applications, including quantum computing, spintronics, and topological quantum technologies. They are also being explored for their potential in realizing new quantum phenomena and for their potential for creating robust, low-dissipation electronic devices. Research in second-generation TIs is a rapidly evolving field, and ongoing efforts are focused on understanding their unique electronic properties, developing new materials with desired properties, and exploring their applications in a wide range of electronic devices [21]. As our understanding of these materials continues to grow, they have the potential to revolutionize various fields of electronics and quantum technologies.

6.3 Higher -order TIs

Higher-order topological insulators are a more recent development in the field of topological materials. These materials exhibit topologically protected modes not only at their surfaces, but also at lower-dimensional boundaries, such as hinges, corners, or edges. These lower-dimensional boundary modes are topologically protected by higher-order bulk topological invariants. Higher-order TIs have been investigated both theoretically and experimentally in multiple systems, including 2D arrays of 1D wires, 3D crystalline materials, and photonic systems, and they have potential applications in areas such as robust quantum information processing and fault-tolerant quantum computing [22]. These materials exhibit topological properties that are dynamically controlled by external time-dependent driving, and they have been experimentally realized in a variety of systems, such as cold atoms, solid-state materials, and photonic systems. Each generation of topological insulators has its own unique properties and potential applications. The field of topological materials is a rapidly evolving area of research, and ongoing studies continue to uncover new types of topological insulators and explore their fundamental physics and practical applications.

6.4 Experimental realization of 2D and 3D TIs

The 2D topological insulator was first presented in 2007. The system consists of HgTe quantum wells squeezed between cadmium telluride. The first 3D topological insulator to be discovered is bismuth antimonide ($Bi_{(1-x)} Sb_x$). Pure bismuth is a semi-metal with a narrow electronic band gap. Found that bismuth antimonide alloys show a strange transition state between a pair of Kramer points and that the mass is characterized by many Dirac fermions. It is also estimated that bulk also contains 3D Dirac particles [7]. This prediction is very interesting because Quantum Hall fractionation has been observed in 2D graphene and pure bismuth. Symmetrically conserved surface states of many Bismuths compounds were also found using ARPES. Many semiconductor materials in the Heusler family are now thought to exhibit topological properties. With the help of doping or grating the Fermi level is made to fall into the gap instead of VB or CB owing to the negative effect of these materials [23].

7. Photonic TIs

A photonic topological insulator is a kind of material that exhibits interesting electromagnetic properties known as topological order, which can lead to robust and highly controllable light propagation behavior. These materials are generally nanoscale structures rather than natural materials. In topological insulators, the behavior of electromagnetic waves (such as light) is determined by the band structure topology, which describes the allowed energy states by electrons or photons in the products. In a conventional insulator, the band structure has a band gap, meaning there are no allowed energy states for electrons or photons within a certain energy range. However, in a topological insulator, the band structure has a negligible topology, even in the presence of defects or impurities. Photonic topological insulators specifically refer to materials where electromagnetic waves, or photons, exhibit topological properties. These materials are designed to manipulate light in unique ways, allowing for novel applications in areas such as optical communication, sensing, and quantum optics [24].

The key attribute of these insulators is the existence of a topologically protected edge state, which is the local type that propagates along the edge of the material without breaking or falling off, even if it is not defective. These edge states can be highly robust and insensitive to external perturbations, making them useful for transmitting information or guiding light in optical circuits with minimal loss or decoherence. Photonic topological insulators can also exhibit other intriguing phenomena, such as the existence of topological defects known as vortices or domain walls, which can trap and manipulate light in unique ways. These properties can be controlled by adjusting the geometry, symmetry or other parameters of the structure. It has the ability to convert many optical devices. By harnessing the unique properties of topological order, photonic topological insulators offer new ways to control light at the Nano scale, with promising applications in fields such as integrated photonics, optical computing, and quantum photonics.

7.1   Floquet topological insulators

Floquet topological insulators belong to quantum many-body systems that exhibit topological properties in periodically driven or time-dependent systems. They are named after the mathematician Gaston Floquet, who studied the behavior of solutions to differential equations under periodic driving. In a conventional topological insulator, the topological properties arise from the intrinsic properties of the system's Hamiltonian, which describes the energy of the system. In Floquet topological insulators, the system is subjected to external time-dependent driving, typically in the form of a periodic electric field or a laser pulse. Under such periodic driving, Floquet topological insulators can exhibit interesting and robust topological properties, even in systems that would otherwise be trivial or non-topological in the absence of driving. These properties are characterized by topological invariants (winding numbers) which can be used to measure the upper properties of Floquet bands or the power spectrum of the driving force [25].

Among the key attributes of Floquet topological insulators, one is that their topological characteristics can be dynamically controlled by adjusting external driving conditions, namely driving frequency or amplitude. This makes Floquet topological insulators a promising platform for realizing topological phenomena in driven quantum systems and for engineering novel quantum states with desired properties. Floquet topological insulators have been demonstrated experimentally in many systems, solid-state materials such as graphene or topological insulator films, and photonic systems. They have potential applications in areas such as quantum information processing, quantum computing, and topological quantum technologies.

8.   Bismuth-based topological insulators

Bismuth-based topological insulators are a type of material that exhibits unique electronic properties, making them interesting for various applications in electronics and quantum technologies. These materials are treated as TI because they have an insulating bulk, which means that they do not conduct electricity through their interior, but they possess conducting surface states that can withstand disorder and impurities. Bismuth-based topological insulators are typically composed of bismuth (Bi) and other elements, such as antimony (Sb), selenium (Se), or tellurium (Te). The most well-known bismuth-based topological insulator is bismuth selenide (Bi_2Se_3), which has been widely studied due to its unique properties. Other bismuth-based topological insulators include bismuth telluride (Bi_2Te_3), bismuth antimony (BiSb), and their alloyed compounds [26]. An important feature of bismuth-based topological insulators is their topologically conserved surface states resulting from the strong spin-orbit coupling of bismuth atoms formed for the following reasons. These surface states exhibit an electric field separation called the Dirac cone, which makes the electrical properties of bismuth-based topological insulators. Bismuth-based topological insulator states are usually located within a narrow energy range, known as the "topological gap", which separates them from the bulk energy bands, making them insulate in the bulk.

Bismuth-based topological insulators have a number of potential applications. For example, their use in spintronic devices where electron spin is used to process and store information is being investigated. Bismuth-based topological insulators are also being explored for use in quantum computing, where their topologically protected surface states could be used to create robust and fault-tolerant cubits for quantum information processing. Additionally, they have potential applications in thermoelectric devices, where they can convert waste heat into electricity. Consequently, bismuth topological insulator-based materials are a special class of surface-protected electronic materials, making them interesting for many applications in electronic, quantum technologies, and energy conversion devices. Ongoing research and development in this area constantly opens up new insights and possibilities for applications for these intriguing materials.

9.   Extensions of one-dimensional topological insulator models

One-dimensional topological insulator (TI) has been extensively studied in recent years for its interesting properties and potential applications in quantum computing and electronics. An extension of the 1D TI model is the SSH model, which includes a dimerization term that distorts the translation of the lattice. This structure exhibits a topological transition with a different dimerization density, and a zero-energy edge state is formed at the transition. These edge states can be extended in one direction and are protected by a topological operation called the wrap number. Another extension of the one-dimensional TI model is the Kitaev model, which includes a superconducting pairing term in addition to the hopping term. This model exhibits a topological phase transition with the emergence of Majorana zero modes at the transition point, as the strength of the pairing term is varied. In addition to the SSH and Kitaev models, many other extensions of the 1D TI model have been explored, including circularly linked models, longitudinally linked models, and more linked models. This model exhibits a wide variety of topological levels and extreme states and has the potential to be used in a variety of quantum technologies.

9.1   SSH model

A 1D lattice model of interacting spin-less fermions that are capable of exhibiting phase transitions as parameters is the Su-Schrieffer-Heeger (SSH) model. It was originally proposed by Su, Schrieffer, and Heeger to describe the electronic properties of a conjugated polymer with alternating single and double bonds known as polyacetylene [27]. The SSH (Su-Schrieffer-Heeger) model is among the most commonly used lattice models to describe topological insulators (TIs). It consists of a chain of sites, where each site has two sublattices labeled "A" and "B". The SSH model exhibits topological properties, such as non-trivial topological edge states, when the parameters t1, t2, and Δ are properly chosen. t1 and t2 are independent jumps between adjacent subgrids A and B where Δ is the energy variation between the A and B sub-lattices, also known as the sub-lattice imbalance or dimerization parameter. It is a simplified model that captures the essential features of some topological insulators, such as the Haldane model and certain organic crystals. The specific values of t1, t2, and Δ depend on the material system under consideration and can be tuned experimentally to realize different topological phases. The SSH model shows a transition point at $\Delta = 0$, where the system moves from a topologically unimportant layer to a non-trivial layer. At a non-trivial level, the system has a nonzero topological constant called the number of turns. Edge particles in the non-elementary phase propagate only in one direction at the edge of the system [28]. It provides a simple and tractable example of phase transition and has potential applications in quantum information processing and electronics.

9.2   Jackiw-Rebbi Model

The Jackiw-Rebbi model is a 1D lattice model of interacting spinless fermions that are capable of exhibiting zero-energy bound states at a single lattice site known as Jackiw-Rebbi zero mode. In 1976 Jackiw and Rebbi presented a simple toy model of soliton dynamics in realistic quantum field theory [29]. This model has a topological origin: The zero mode is the result of the product of the model defined by a topological invariant called the winding number. If the winding number is not zero, then there is a connected state of zero energy at the center of the lattice arising from the topological protection of the state. The Jackiw-Rebbi model has been widely studied in condensed matter physics as well as in TIs and superconductors. It provides simple and easy examples of topologically connected states with applications in quantum computing and electronics.

10.   Reversed conductance decay of 1D topological insulators

The reverse conduction decay of 1D topological insulators is a phenomenon in which the increment in the disorder of topological insulators causes an increment in its conductivity of topologically protected edge state in a 1D lattice system as compared with the normal behavior of conductance decay in disorder systems. This happens because of weak topologically conserved states which remain delocalized even if they encounter an impurity. The topologically conserved state in a 1D topological insulator is a chiral state propagating along with the boundary of two systems with different invariants. The conductivity of the edge state is calculated in units of conductivity. However, when the power conflict rises above a certain threshold, states become localized, and their practices disappear with the length of the system. This behavior is typical of Anderson localization, where the disturbance causes an interaction between transport and local electronic states [30].

Conversely, in the presence of conflict, nation-state behavior may backfire, where behavior increases with the conflict. This effect is due to the interaction between the disorder because of scattering and the topologically protected material at the boundary of the state. In particular, edge states can be distributed but still remain localized due to their topological protection. Thus, as the strength of the conflict grows, the edge of the state can gain more structure and become more effective. Inverse decay has been observed in many studies and calculations of 1-dimensional topological insulators, including Su-Schrieffer-Heeger-based models and Kitaev chains. This effect has the potential to affect the creation of topological devices however it should be noted that reverse conductivity degradation is a negative effect depending on the context of the system and the nature of the problem. In addition, reverse conduction decay is only observed in strong enough conditions, where the weak current causes the static behavior to resume. Therefore, reverse conductivity degradation should be understood as a special feature of some 1-D topological insulators rather than a general feature of conflicting systems.

11. Topological Insulators in a ten-fold way

The tenfold way is a classification scheme for topological insulators based on their symmetry properties. It was proposed in 2010 by Schnyder, Ryu, Ludwig and Furusaki and is established on the symmetry classification of Altland and Zirnbauer. The tenfold way classifies topological insulators into ten distinct classes, which are distinguished by their dimensionality, symmetry properties, and the number of protected topological invariants. The tenfold method divides topological insulators into ten classes, distinguished by their dimensionality, symmetry properties, and number while retaining topological invariants. These classes are represented by the letters A, AII, AIII, BDI, D, DIII, CI, CII, CIII, and AIII and are divided into five classes, A, AII, AIII, BDI, and III. The presence of certain symmetries is responsible for the determination of angular equilibrium. These symmetries preclude the existence of topologically conserved states that do not have significant topological invariants at the boundary, like Chern numbers, Z_2 invariants, and wrapping numbers [31].

In particular, the tenfold way provides a unified framework for understanding the topological properties of various types of topological insulators, including 1D systems like the SSH model, 2D systems such as the QHS, and 3D systems such as the topological insulators in bismuth telluride and other materials. The tenfold way has been widely used in both theoretical and experimental studies of topological insulators and opens a gateway for many new topological phases of matter. Overall, the tenfold way provides a powerful tool for understanding the rich variety of topological phenomena that can arise in condensed matter systems and has stimulated much research in this area over the past decade. These are some of the common ways to classify TIs based on symmetries. It's worth noting that different TIs may exhibit combinations of these symmetries, and their classifications may depend on the specific details of their band structures and crystal symmetries. Moreover, it is a rapidly evolving field, and ongoing research continues to uncover new insights into their classification and properties.

11.1 T-symmetry

An important symmetry of the physical system that describes the behavior of reversed time systems is the T-symmetry or Time reversal symmetry. It states that if all the directions of motion of particles in a physical system are reversed, the resulting system should be physically indistinguishable from the original system. Mathematically, it is denoted as T and defined such that $T^2 = +1$. It reverses the direction of momentum and time when applied to a wave function describing a physical system. Usually, the existence of topological states in certain materials, for instance, topological insulators, can be protected by this symmetry which is identified by nontrivial topological invariants like Z_2 invariant [32]. For instance, the existence of a pair of back-propagating edge states protected from backscattering by a Z_2 invariant at the system boundary is guaranteed by T-symmetry in the absence of this inverse symmetry the boundary will be affected by the presence of impurities or other conflicts. Observations of time reverse symmetry-protected topological insulators in various materials such as HgTe/CdTe quantum wells, bismuth selenide (Bi_2Se_3) and bismuth telluride (Bi_2Te_3) have been reported. In general, T-symmetry is an essential concept in physics, especially in a topological context, which is an important factor to determine the physical behavior of insulators and other topological materials.

11.2   Particle-hole symmetry

A rudimentary symmetry of physical systems that describes the behavior of a system under the exchange of particles with holes is Particle-hole symmetry. It states that if all the particles in a physical system are replaced with holes (and vice versa), the resulting system should be physically indistinguishable from the original system. When applied to a wave function describing a physical body, the particle-hole operator is denoted as C, which is defined such that $C^2 = +1$. Changes the signs of each transition and replaces them with the corresponding holes. In condensed matter physics, electronic properties are often determined by particle-hole symmetry especially those with superconductivity. In addition to its role in superconductivity particle-hole symmetry can also protect the existence of topologically protected states in certain materials, such as topological insulators and topological superconductors. Particle-hole symmetry in general is an important concept in physics that is an important factor in the determination of topological and electronic properties, especially in the element language of superconducting and topological materials.

11.3   Chiral symmetry

Chiral symmetry is an important symmetry of the physical body and describes how the body behaves when left-handed particles are replaced by right-handed particles (or vice versa). It is also known as handedness or helicity symmetry. Mathematically, chiral symmetry is represented by an operator denoted as $\gamma 5$, which is defined such that $\gamma 5^2 = +1$. When applied to a wave function describing a physical system, this operator reverses the sign of the momentum of left-handed particles and the sign of the helicity (i.e., spin projection along the momentum direction) of right-handed particles. In particle physics, the weak interaction of one of the four forces of nature can be explained by the Chiral symmetries. The weak interaction is known to violate parity symmetry (the symmetry under spatial inversion) and charge conjugation symmetry (the symmetry under particle-antiparticle exchange), but it was initially thought to conserve chiral symmetry [33]. However, it was later discovered that chiral symmetry is also violated by weak interaction, albeit to a much smaller extent than parity and charge conjugation. Chiral fermions described the electronic states near the Fermi level, which have a definite handedness and are protected by chiral symmetry. Overall, chiral symmetry is a fundamental concept in physics that describes both particle physics and condensed matter physics, particularly the weak interaction in topological materials.

12. Future evolution of 1D topological insulators

As discussed above, one-dimensional topological insulators belong to a special kind of material that exhibit unique electronic properties, where the edges of the material behave as conducting channels while the bulk remains insulating [34]. Here are some possible future evolutions in the field of one-dimensional topological insulators:

Innovation of new materials: One-dimensional topological insulators have been primarily studied in materials such as nanowires, Nano ribbons, and atomic chains. In the future, there could be the discovery of new materials that exhibit one-dimensional topological insulator behavior, expanding the range of materials available for practical applications. This could involve exploring unconventional materials or heterostructures with tailored properties to enhance the performance of one-dimensional topological insulators.

Fine-tuning of electronic properties: Researchers could develop methods to precisely control the electronic properties of one-dimensional topological insulators, such as their band gap, Fermi level, and edge states. This could be achieved through various approaches, including strain engineering, chemical doping, or electric field gating. Fine-tuning the electrical properties of 1D topological insulators can help design devices that match the desired performance.

Integration with other materials and systems: One-dimensional topological insulators could be integrated with other materials or systems to create hybrid structures with enhanced properties. For example, coupling one-dimensional topological insulators with superconductors could lead to the realization of topological superconductors, which are exotic quantum states with potential applications in fault-tolerant quantum computing. Integration with other materials and systems could also enable the development of novel devices, such as topological transistors, sensors, and memory devices.

Exploration of new physical phenomena: The unique properties of 1-dimensional topological insulators are a gateway to the discovery of the latest physical phenomena. For example, interactions between topological edge states and electron-electron interactions are responsible for phenomena such as fractional edge states, Majorana-dependent states, and non-Abelian anions. These quantum particles could potentially be used in quantum information processing, quantum communication, and quantum sensing.

Technological advancements: Advances in fabrication techniques, characterization methods, and computational simulations could significantly impact the field of one-dimensional topological insulators. For instance, the development of advanced nanofabrication techniques could enable the fabrication of complex one-dimensional topological insulator structures with precise control over their properties. Furthermore, computational simulations could aid in predicting the properties of novel one-dimensional topological insulators and guide experimental efforts.

Real-world applications: One-dimensional topological insulators have the potential for various real-world applications. For instance, they could be used in low-power electronic devices, where the edge states could enable highly efficient charge transport with reduced dissipation. They could also be employed in spintronic devices, where the spin-momentum locking property of edge states could be harnessed for spin manipulation and information processing. Additionally, one-dimensional topological insulators could find applications in quantum computing and quantum communication, as their topological properties could provide robustness against decoherence and error correction.

The field of one-dimensional topological insulators is a vigorous area of research with promising prospects for future developments. These could include the discovery of new materials, fine-tuning of electronic properties, integration with other materials and systems, exploration of new physical phenomena, technological advancements, and real-world applications. Continued research in this field is expected to yield exciting breakthroughs with potential applications in a wide range of fields.

Conclusion

Topological insulators are another important discovery that will certainly drive the expansion of science. 1D topological insulators are a particular genre of materials that exhibit special electrical behavior due to their topological nature. These materials often have a chain-like or thread-like pattern that makes them look the same. They do not conduct electricity through their interiors while exhibiting conducting states at their surfaces or edges that are protected by topology. Topological order is responsible for the formation of energy bands with nontrivial topology, resulting in topologically protected surface or edge states that are robust against impurities and disorder. Both dimensions of the material and its elemental symmetries tell us about the premises of topological insulators along with its

Materials Research Forum LLC
https://doi.org/10.21741/9781644902851-2

surface states Edge states are characterized by their unique dispersion relations, which are "spin-momentum locked". This makes them robust against scattering and backscattering, making them ideal for spintronic applications. The bulk of topological insulators remains insulating, and with a band gap that prevents electrical conduction in the bulk. This property makes them distinct from ordinary conductors, where conduction occurs throughout the material. Topological insulators can exhibit unique quantum phenomena, such as QHS, which is a quantum analog of the classical Hall Effect, but with spin degrees of freedom playing a crucial role. Because of their unique electronic properties, topological insulators have dormant applications in a wide range of fields, including spintronics, quantum computing, and topological quantum computing. Overall, TIs are a fascinating group of quantum materials that give rise to great interest in both fundamental research and technological applications, with the promise of revolutionizing various fields of modern electronics and quantum technologies.

References

[1] J. Moore, The next generation, Nature Physics 5 (2009) 378-380. https://doi.org/10.1038/nphys1294

[2] B.A. Bernevig, S.C. Zhang, Quantum spin Hall effect, Physical Review Letters 96 (2006) 106802. https://doi.org/10.1103/PhysRevLett.96.106802

[3] H. Beidenkopf, P. Roushan, J. Seo, L. Gorman, I. Drozdov, Y.S. Hor, R.J. Cava, A. Yazdani, Spatial fluctuations of helical Dirac fermions on the surface of topological insulators, Nature Physics 7 (2011) 939-943. https://doi.org/10.1038/nphys2108

[4] Z. Zhu, Y. Cheng, U. Schwingenschlögl, Band inversion mechanism in topological insulators: A guideline for materials design, Physical Review B 85 (2012) 235401. https://doi.org/10.1103/PhysRevB.85.235401

[5] X.L. Qi, S.C. Zhang, Topological insulators and superconductors, Reviews of Modern Physics 83 (2011) 1057. https://doi.org/10.1103/RevModPhys.83.1057

[6] O. Pankratov, Electronic properties of band-inverted heterojunctions: Supersymmetry in narrow-gap semiconductors, Semiconductor Science and Technology 5 (1990) S204. https://doi.org/10.1088/0268-1242/5/3S/045

[7] M. Konig, S. Wiedmann, C. Brune, A. Roth, H. Buhmann, L.W. Molenkamp, X.L. Qi, S.C. Zhang, Quantum spin Hall insulator state in HgTe quantum wells, Science 318 (2007) 766-770. https://doi.org/10.1126/science.1148047

Materials Research Forum LLC
https://doi.org/10.21741/9781644902851-2

[8] W. Tian, W. Yu, J. Shi, Y. Wang, The property, preparation, and application of topological insulators: A review, Materials 10 (2017) 814. https://doi.org/10.3390/ma10070814

[9] D. Hsieh, Y. Xia, D. Qian, L. Wray, J. Dil, F. Meier, J. Osterwalder, L. Patthey, J. Checkelsky, N.P. Ong, A tunable topological insulator in the spin helical Dirac transport regime, Nature 460 (2009) 1101-1105. https://doi.org/10.1038/nature08234

[10] X.L. Qi, S.C. Zhang, The quantum spin Hall effect and topological insulators, arXiv preprint arXiv:1001.1602 (2010).

[11] A.P. Schnyder, S. Ryu, A. Furusaki, A.W. Ludwig, Classification of topological insulators and superconductors, AIP Conference Proceedings, American Institute of Physics, pp. 10-21, 2009. https://doi.org/10.1063/1.3149481

[12] Y. Xue, H. Huan, B. Zhao, Y. Luo, Z. Zhang, Z. Yang, Higher-order topological insulators in two-dimensional Dirac materials, Physical Review Research 3 (2021) L042044. https://doi.org/10.1103/PhysRevResearch.3.L042044

[13] D. Kong, J.C. Randel, H. Peng, J.J. Cha, S. Meister, K. Lai, Y. Chen, Z.X. Shen, H.C. Manoharan, Y. Cui, Topological insulator nanowires and nanoribbons, Nano Letters 10 (2010) 329-333. https://doi.org/10.1021/nl903663a

[14] S.S. Hong, W. Kundhikanjana, J.J. Cha, K. Lai, D. Kong, S. Meister, M.A. Kelly, Z.X. Shen, Y. Cui, Ultrathin topological insulator Bi2Se3 nanoribbons exfoliated by atomic force microscopy, Nano Letters 10 (2010) 3118-3122. https://doi.org/10.1021/nl101884h

[15] X. Chen, X.C. Ma, K. He, J.F. Jia, Q.K. Xue, Molecular beam epitaxial growth of topological insulators, Advanced Materials 23 (2011) 1162-1165. https://doi.org/10.1002/adma.201003855

[16] M. Liu, F.Y. Liu, B.Y. Man, D. Bi, X.Y. Xu, Multi-layered nanostructure Bi2Se3 grown by chemical vapor deposition in the selenium-rich atmosphere, Applied surface science 317 (2014) 257-261. https://doi.org/10.1016/j.apsusc.2014.08.103

[17] V.S. Stolyarov, D.S. Yakovlev, S.N. Kozlov, O.V. Skryabina, D.S. Lvov, A.I. Gumarov, O.V. Emelyanova, P.S. Dzhumaev, I.V. Shchetinin, R.A. Hovhannisyan, Josephson current mediated by ballistic topological states in Bi2Te2.3Se0.7 single nanocrystals, Communications Materials 1 (2020) 38. https://doi.org/10.1038/s43246-020-0037-y

[18] S. Panahiyan, S. Fritzsche, Toward simulation of topological phenomena with one-, two-, and three-dimensional quantum walks, Physical Review A 103 (2021) 012201. https://doi.org/10.1103/PhysRevA.103.012201

[19] W. Zhang, R. Yu, H.J. Zhang, X. Dai, Z. Fang, First-principles studies of the three-dimensional strong topological insulators Bi2Te3, Bi2Se3 and Sb2Te3, New Journal of Physics 12 (2010) 065013. https://doi.org/10.1088/1367-2630/12/6/065013

[20] D. Hsieh, Y. Xia, D. Qian, L. Wray, F. Meier, J. Dil, J. Osterwalder, L. Patthey, A. Fedorov, H. Lin, Observation of time-reversal-protected single-Dirac-cone topological-insulator states in Bi2Te3 and Sb2Te3, Physical Review Letters 103 (2009) 146401. https://doi.org/10.1103/PhysRevLett.103.146401

[21] M.Z. Hasan, J.E. Moore, Three-dimensional topological insulators, Annual Review Condensed Matter Physics 2 (2011) 55-78. https://doi.org/10.1146/annurev-conmatphys-062910-140432

[22] Y. Ando, Topological insulator materials, Journal of the Physical Society of Japan 82 (2013) 102001. https://doi.org/10.7566/JPSJ.82.102001

[23] H.J. Noh, H. Koh, S.J. Oh, J.H. Park, H.D. Kim, J. Rameau, T. Valla, T. Kidd, P. Johnson, Y. Hu, Spin-orbit interaction effect in the electronic structure of Bi2Te3 observed by angle-resolved photoemission spectroscopy, Europhysics Letters 81 (2008) 57006. https://doi.org/10.1209/0295-5075/81/57006

[24] M. Geier, L. Trifunovic, M. Hoskam, P.W. Brouwer, Second-order topological insulators and superconductors with an order-two crystalline symmetry, Physical Review B 97 (2018) 205135. https://doi.org/10.1103/PhysRevB.97.205135

[25] M. Khazali, Discrete-time quantum-walk & Floquet topological insulators via distance-selective Rydberg-interaction, Quantum 6 (2022) 664. https://doi.org/10.22331/q-2022-03-03-664

[26] S.T. Pi, H. Wang, J. Kim, R. Wu, Y.K. Wang, C.K. Lu, New class of 3D topological insulator in double perovskite, The Journal of Physical Chemistry Letters 8 (2017) 332-339. https://doi.org/10.1021/acs.jpclett.6b02860

[27] W.P. Su, J. Schrieffer, A.J. Heeger, Solitons in polyacetylene, Physical Review Letters 42 (1979) 1698. https://doi.org/10.1103/PhysRevLett.42.1698

[28] R. Shankar, Topological insulators-A review, arXiv preprint arXiv:1804.06471 (2018).

[29] R. Jackiw, C. Rebbi, Solitons with fermion number ½, Physical Review D 13 (1976) 3398. https://doi.org/10.1103/PhysRevD.13.3398

[30] L. Li, S. Gunasekaran, Y. Wei, C. Nuckolls, L. Venkataraman, Reversed conductance decay of 1D topological insulators by tight-binding analysis, The Journal of Physical Chemistry Letters 13 (2022) 9703-9710. https://doi.org/10.1021/acs.jpclett.2c02812

[31] A.P. Schnyder, S. Ryu, A. Furusaki, A.W. Ludwig, Classification of topological insulators and superconductors in three spatial dimensions, Physical Review B 78 (2008) 195125. https://doi.org/10.1103/PhysRevB.78.195125

[32] A.W. Ludwig, Topological phases: Classification of topological insulators and superconductors of non-interacting fermions, and beyond, Physica Scripta 2016 (2015) 014001. https://doi.org/10.1088/0031-8949/2015/T168/014001

[33] M.Z. Hasan, C.L. Kane, Colloquium: Topological insulators, Reviews of Modern Physics, 82 (2010) 3045. https://doi.org/10.1103/RevModPhys.82.3045

[34] J.K. Asbóth, L. Oroszlány, A. Pályi, A short course on topological insulators, Lecture notes in Physics, 2016. https://doi.org/10.1007/978-3-319-25607-8

Topological Insulators: Materials and Applications   Materials Research Forum LLC
Materials Research Foundations 154 (2024) 47-60   https://doi.org/10.21741/9781644902851-3

Chapter 3

The Origin of Topological Insulators

Maria Wasim[1*], Aneela Sabir[1], Muhammad Shafiq[1], Rafi Ullah Khan[1]

[1]Institute of Polymer and Textile Engineering, University of the Punjab, Lahore, 54590 Pakistan

*maria-be24@hotmail.co.uk, mariawasim.ipte@pu.edu.pk

Abstract

On the surfaces of several insulators, unusual metallic states exist. These states are created by topological phenomena, which also make the movement of electrons over interfaces that are impervious to impurity scattering. These types of topographical insulators could open up innovative pathways for creating novel phases and particles, which might find value in spintronics and quantum computing applications.

Keywords

Topological Insulator, Physics, Graphene

Contents

1. Introduction

Understanding how order develops when an extremely high number of basic components, including such electrons, ions, or the magnetic moment, interrelate with one another is a major focus of condensed-matter physics. As ions are distributed regularly in crystals due to their electrostatic contacts, the uninterrupted space symmetry throughout translational and rotational movement is broken. Similarly, in conventional magnets, certain aspects of the symmetry of rotational movement in space are destroyed along with the symmetry at time reversal.

The topological sort of order that underlies the quantum Hall effect is shown by electrons that are constrained to 2D and exposed to a powerful magnetic field, which was a significant finding in the 1980s. Dissipation-less transport, emerging particles with fractional-based charge, and statistics are some effects of this order. One of the significant findings in recent years is the occurrence of topological order in certain 3D materials. In these materials, the inherent feature of all solids known as spin-orbit coupling, which plays the part of the magnetic field, assumes the function of the magnetic field. Topological insulators are materials that are insulators in large quantities but exhibit strange metallic phases at their interface as a result of the topological order.

We give an outline of the fundamental ideas driving topological insulators and ongoing research on these amazing new materials in this perspective chapter. Following a transitory past of such a swiftly evolving arena and a clarification of what tends to make some insulators "topological," We discuss recent developments in experimental studies on topological insulators, with together mass and nanostructured materials, as well as in the conceptual underpinnings of such materials. We finish by describing why several research teams are attempting to create novel particles and phases using topological insulators, which might have applications to quantum computers.

2. Topological insulator's primer

An insulator that consistently has a metallic border as compared to a vacuum or a "regular" insulator is the simplest method to define a topological insulator. These topological attributes, which cannot alter as long as a substance is insulating, are the source of these metallic barriers. As shown in Fig 1, in order to demonstrate a closed system and topological insulator, the trefoil knot is needed to symbolize a conventional one, providing an understandable explanation of why metallic surfaces exist. The study of qualities of things that remain unchanged when subjected to smooth deformations is known as topology. A typical example is the transformation of a doughnut into a coffee cup. Closed loop along with the trefoil knot, in comparison to the doughnut and cup of coffee combination, have unique topological invariants. As a result, no regardless of how the thread (or wire) is twisted or stretched without being cut, neither can be distorted to become the other. The outer surface cannot continue being an insulating nature, which is analogous to severing tie knot because these invariants do alter while traversing the border between topological and conventional insulators.

Figure.1 (a) Simple loop, (b) Trefoil knot, and (c) knotted 3D electronic structure (Hopf map)

The element that is "knotted" in a topological insulator is considered to be the electron wave function as it would travel via the momentum space. Such knotting phenomena are accompanied by various topological factors, which are constants as long as the substance is insulating (often stated as integrals using the wave function). There exist metallic delocalized wave functions at the borderline between conventional and topological insulators, even though this barrier may appear to defy the continuum description using topology. A modest example of knot phenomena in a three-dimensional electronic structure is given in Fig 1c. For one electronic band that is occupied and one that is unoccupied, an individual point in three-dimensional momentum space is connected to a unit vector that represents the phase that is occupied, while the Hopf map is presented in Fig. 1c.

2.1 Knowledge acquire from past

The primary experimental indicator that such an insulator is in fact topological is the emergence of the peculiar metal whenever the topological fluctuations occur at the surface. We will briefly recap the chronological processes that lead to the model expectations that topological insulators emerge before going on to discuss the features of the insulator. In a quantum Hall droplet, which is the initial known instance 2D topological order, a simpler variation of this metal may be found near the edge. Quantum Hall edges, in which the insulating droplets are tightly wrapped with quantum wires (Fig. 2I), however when the electrons of the components are restricted to only 2 dimensionality and the application of robust electromagnetic field in the direction opposite to which they are bound. This is due to the topological properties of electronic wave functions [1].

With the idea that quantum Hall phenomena can be seen in electron species while they are in motion in lattice structure regardless of any magnetic field on a macroscopic level, the topological study was sparked. In the late 1980s, it was suggested that rather than being guided by such a magnetic field, electrons theoretically produce a state of quantum Hall that is controlled by the forces generated because of the motion across the lattice structure of the crystal. The basis of current advancement is the relativistic phenomena known as spin-orbit coupling, which relates the freedoms granted for the orbital and spin-based angular momentum of electron species. Even in non-magnetic materials, this connection causes electrons to encounter the force governed by spin as they move along the crystal structure.

Due to the violation of time reversal symmetry in applied magnetic field, coupling of orbit-spin lacks the balance to initiate the phenomena of quantum Hall. The phenomena of quantum spin Hall may happened when electron species having opposite spin angular momentum (spin up and spin down) move in reverse channel around the rim of the droplet devoid of any imposed magnetic field, can nevertheless be produced by spin-orbit coupling in simplified models that were first suggested in roughly 2003 [2] (Fig. 2II). Using such more basic models, the foremost stepladders concerning comprehending topological insulators were taken. As the electron having the spin-up and spin-down movement commonly mix in real materials and there is no such preserved spin current, it was unclear how realistic the models were. It was also uncertain if the drop in Fig. 2II edge condition would endure the presence of even a small number of contaminants.

Kane and Mele [3] achieved a significant theoretical advancement in 2005. They demonstrated that certain aspects of quantum spin Hall phenomena can endure using more realistic models that did not include a preserved spin current. They discovered a brand-new kind of topological type that, when calculated for any two D material, could tell whether or not it possessed a stable edge state. The resulting 2-dimensional state became the initial topological insulator to also be understood, and it certify to demonstrate that, even though the edge has not been stabilize enough in existing models, there exist genuine 2 dimensional materials which would possess a stabilized edge phase having zero magnetic field. At low temperatures, edges of this non-magnetic insulator behave as quantum Hall effect-like completely conducting one-dimensional electronic wires.

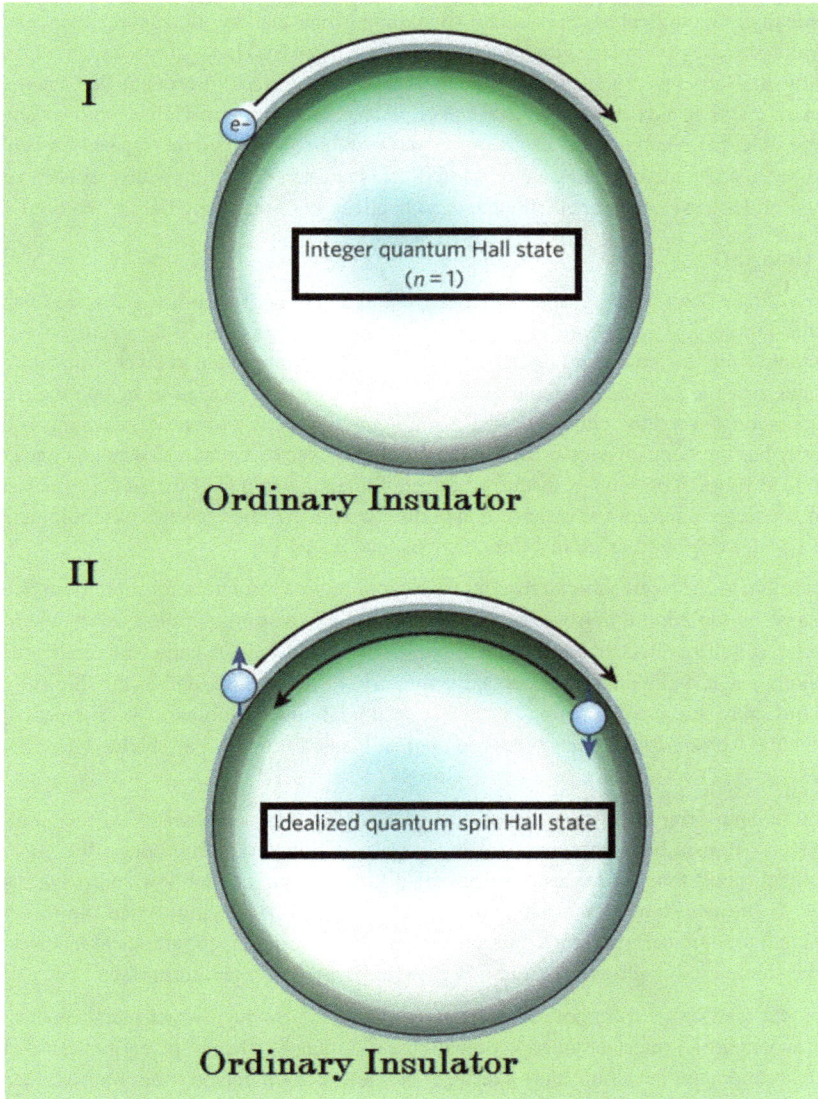

Figure 2. (I) Topographical order in two dimensions.

Theoretically, topological based insulator (two dimensional) having fixed conductance of charge alongside the boundaries may eventually be achieved in (Hg,Cd)Te quantum wells, according to Bernevig, Hughes, and Zhang [4]. In this system, the quantified charge conductance was in fact noticed in 2007 as a plateau like quantum-Hall devoid of any magnetic field. [5]. Such researches are comparable to those on the effect of quantum Hall in that they need low temperatures and synthetic 2D materials (quantum wells), at least up to this point. However, they are distinct as zero application of magnetic field is required.

2.2   Going 3D

Most significant theoretical advancement occurred in 2006 [6-8] when it was realized that while the topological insulator would, in ingenious sense, generalize to a really three dimensional state, the quantum Hall effect doesn't really. Comparable to layered quantum Hall states, a weak three-dimensional topological insulator may be created by coating the two-dimensional versions, however the resulting state is not disorder-resistant and essentially has the same physics as the 2D state. The presence of a quantum wire analogous to the one at verge of previously discussed effect of quantum spin Hall would always exist in weak topological insulators, which could permit two dimensional physics of topological based insulator to be witnessed in a three dimensional matter [9].

It has feasible to link conventional insulators and topological insulators effortlessly in two dimensions by violating time-reversal symmetry [7]. Such a strong topological insulator, however, has quite a nuanced connection to two-dimensional case. A band structure (three dimensional) that maintains time-reversal symmetry, is not coated, and is topologically non-trivial may be created using such a continuous interpolation. The subject of experimental work has been this potent topological insulator, which has shielded metallic surfaces.

Once more, spin-orbit coupling is necessary, and its necessity combines all of the spin's components. It can be comprehended as, there is no method to produce the three dimensional robust topological insulator having distinctive spin up and down electrons, as opposed to the two-dimensional situation. This creates it challenging to see the three-dimensional topological insulator's bulk physics (solely the robust topographical insulator would be studied moving forward), but it is easy to visualize its metallic surface.

Topological features as from core insulator are "inherited" by the distinct metal (planar) that cultivates at the interface of topological based insulator. On the plane interface, in which the momentum along the interface is clearly demarcated, the bulk-surface connection manifests itself in its most basic form. Each momentum all along surface has sole state of spin only at the Fermi level, and the direction of spinning changes with the change in momentum that cover all around Fermi interface. Scattering in between surface states will occur when disorder or contaminants are incorporated to the interface, but importantly, the topological characteristics of the core insulator prevent the metallic result indicates from disappearing or becoming localized or gapped. In the past two years, there has been an explosion of experiments conducted on 3D topological insulators as a result of these twin theoretical expectations concerning the electrical structure of the substrate interface and the resistance to chaos of its metallic behavior.

3. Experimental realizations

The alloy Bi_xSb_{1-x} was the first topological insulator to be found, and angle-resolved photoemission spectroscopy (ARPES) experiments were used to map its peculiar surface bands [10, 11]. In ARPES experiments, an electron is ejected from a crystal by a high-energy photon, and the interface or core electronic structure is subsequently identified by examination of the electron's ejected momentum. Despite the complicated surface morphology of this alloy being discovered, this finding started the hunt for new topological insulators.

The most likely possibilities are heavy-element, small-bandgap semiconductors because of coupling of orbit-spin needs tends to be robust enough to drastically alter the structure (electronic) in order for a topological insulator to emerge. This concept is based on two ideas. First, the relativistic phenomenon known as spin-orbit coupling only occurs strongly for heavy materials. Secondly, the coupling of orbit-spin cannot guarantee the modification if the bandgap is considerably bigger than that of the scale of energy of coupled orbit-spin. Recently, topological insulator behavior in Bi_2Se_3 and Bi_2Te_3 was discovered, which was the culmination of the hunt for topological insulators [12-14].

These "next-generation" materials have the simplest permitted surface state and showed the properties of topological insulator at higher temperature range than the original form of material (Bi_xSb_{1-x}). They consist of bulk bandgaps (greater than 0.1 eV). The detail explanation of such material opens the doors of numerous application few of them are mentioned here, moreover, it supports the hypothesis involving the topological insulator. Additionally, these tests are not required to be performed at low temperatures due to the enormous bandgap. The biggest continuing issue with these materials is the presence of contamination in bulk form as a result of residual conductivity. This is especially problematic when utilizing experimental methods that do not separate the bulk form or the interface phase.

3.1 A graphene lookalike

Dirac based structure (electronic) of graphene molecule which is also known as a Dirac cone. It possesses a linear based energy momentum, a connection comparable to the particle (relativistic), is intimately connected to the superficial state of the subsequent age topological insulators. Research on graphene, which is the most 2D substance that is feasible and possesses a linear energy-momentum connection, has been quite active in recent years [15]. It is intriguing from both a structural and an electrical perspective. The topological insulator's surface differs significantly from graphene's in that it contains two points of Dirac and spin can be degenerate, compared to the topological insulator's single point of Dirac and lack of ability of spin to degenerate. This distinction has profound effects, including the potential for the creation of novel particles with potential uses in quantum computing.

Scanning tunnelling microscopy experiments [16-18] revealed another striking result of the absence of additional degeneracies: interference patterns near surface defects or steps demonstrate that scattered electrons are never entirely reflected. The surface is still metallic [19], even if the disorder intensifies significantly and a description based on clearly distinct scattering events is erroneous. In some cases like in noble metals, the dissimilarity among the interface of topological insulator and the accidental interface states present is due to the defence phenomena occuring on the interface of metals due to the Anderson localization, which causes the formation of insulating layer due to the presence of strong disorder [20]. It is well known that the topological based insulator, at some point particularly in few 2 dimensional based models, arises as a result of disorder, in addition to being stable to disorder [21, 22]. If the dispersion has an extremely smooth potential, graphene can mimic this protection, but genuine graphene having severe disorders are more likely to be localized.

However, there might be some demerits for the usage of topological based insulator in contrast to graphene for certain applications. Fermi level is the point where the density of any state tends to disappear at Dirac point, at which both cones cross each other, because of the chemical structure of carbon atoms present in graphene. That cause the usage of graphene in both fundamental and micro-electronics research due to the flexibility of the density of carriers present in graphene that can be used to generate the electric field. Although a topological insulator's surface Fermi level has no special purpose to be present at Dirac point, for the case of Bi_2Se_3, changing this point has lately shown by the coupling of chemical modification of bulk and interface [23]. For the production and use of topological based exciton condensate via the biasing a skinny layer of topological insulator both rely on this regulation of chemical potential [24]. Such exciton-based condensate is an electron-hole superfluid phase with bound electronic state that surrounds superfluid vortices.

3.2 Concerned matter

Binary contemporary advancements in the processes used to create of topological insulators are also significant to mention. The effect of Aharonov-Bohm can be detected in the interface of metals, whenever a magnetic field is generated to the length wise dimension of nanoribbon [25] thanks to the use of molecular beam epitaxy to manufacture films of Bi_2Se_3 with adjustable width to just one unit cell [26]. Several scientists suggested uses of topological based insulators in the field of spintronics and further domains depend on these nanostructures. For the magnetic based memory application, the novel class of spin-based torque gadget would be possible using a topological insulator and ferromagnet heterostructure. When the current flows via the exterior of topological based insulator, ferromagnet can be switched. The next crucial stages involve measuring the conductivity and spin characteristics of the metallic substrate surface using straight transport and optical tests. Better materials can be employed for these trials that composed of condensed residual bulk conductivity [27].

4. A novel field

Numerous ongoing investigations are established on an alternative interpretation of the fundamental characteristics of a topological based insulator, an interpretation linked to the late nineteen mainly for the research on particle physics. It has been understood that in the interior of certain insulators, the fields of electric and magnetic are related. The linking of electric and magnetic fled to the axion based was being investigated by particle physicists, first explored a unique kind of quantized coupling that is present in topological insulators.

4.1 Superfluidity and particle physics

Condensed matter physics researchers frequently try to describe a state of matter in terms of how it reacts to an external force. For instance, a superconductor is identified by means of the Meissner effect (via the evacuation of enforced magnetic field), but a solid material is described having non-0 stiffness in reaction to shear force applied. Such definitions do not depend on microscopic details and are directly related to experimentation. It turned out that the description of reaction of a topological based insulator was nearly finished in late nineteen in an determination to comprehend the attributes of axion based electro-dynamics, which distinguish from regular electro-dynamics by adding to the Lagrangian a word means the magnetic and electric fled are directly proportional to each other or the scalar product (E.B) [28]. Just twin values of the coefficient of this terminology, which correspond to conventional and topological insulators, are consistent with time-reversal symmetry. As a result, topological insulators are substances whose interior configuration produces a non-0 value for the axion-like coupling, much to how insulators alter the dielectric constant, resulting in the Lagrangian's coefficient E^2 [29].

This term is expected to have a wide range of effects, including surface states with varying Hall conductivity and monopole-like behavior [30, 31]. The straightforward magnetoelectric effect, whereby an imposed electrical field produces a magnetic dipole and conversely, may be the most significant for applications. Since the magnetoelectric impact in topological insulators arises only from electron orbital motion, it has the potential to be more rapid and reproducible without fatigue than the magneto-electric effect in multiferroic materials [32]. Simply asking how much a given material's bulk polarization shifts in a magnetic field can theoretically determine whether it is a topological insulator, and evaluation of this magneto-electric polarizability in those further materials imply that it ought to be feasible to notice the impact in topological insulators.

4.2   Emergent particles and quantum computing

Emergent particle that none of the material can support on its own may be created at the contact in between topological insulator and a superconductor. When a topological insulator is put adjacent to a regular superconductor, the proximity effect causes the metallic surface to become superconducting like any other metal. A 0 energy Majorana fermion is confined around the vortex core in the region of the super conducting coating of a topological insulator, however, on a condition whether a line of vortex proceeds to the insulator via the superconductor [33, 34]. Majorana fermion possesses quantum numbers that are different in contrast to normal electron as they have their own antiparticle which is neutral (electrically), and usually, "half" of normal electron [35]. This contrasts with vortex core states in regular superconductors. The anticipation around this plan and others with a comparable spirit is caused by a number of factors [36]. First, these ideas could make it possible to directly see a Majorana fermion, a long pursued after objective mainly in condensed matter and particle physics. Secondly, Majorana fermions are the primarily step in the direction of topological quantum computer, that might be extraordinarily error-free because the quasiparticles follow a unique form of non-Abelian quantum statistics [37]. The variety of topological phenomena that have been witnessed in heterostructures based semiconductor mainly in the last 30 years followed one to believe that topological based insulators which were previously discovered are merely the beginning and that additional types of topological materials of topological order are yet to be discovered.

Conclusion

Similar to other recent developments in fundamental condensed-matter physics, the revelation of topological insulators opens the door to novel applications that expand on our newly acquired knowledge. These insulators' peculiar metallic surfaces might lead to brand-new spintronic or magneto-electric devices. Topological insulators might also result in a novel topological quantum bit architecture when combined with superconductors. These insulators have already significantly influenced the physics of pure condensed matter, demonstrating that topological impact which was considered previously to be restricted to relatively low temperature, condensed dimension, as well as the elevated magnetic fields, may affect the dynamics of majority of the materials that appear to be ordinary under normal circumstances.

References

[1] F.D.M. Haldane, Model for a quantum hall effect without Landau levels: Condensed-matter realization of the "parity anomaly", Physical Review Letters 61 (1988) 2015-2018. https://doi.org/10.1103/PhysRevLett.61.2015

[2] S. Murakami, N. Nagaosa, S.C. Zhang, Spin-hall insulator, Physical Review Letters 93 (2004) 156804. https://doi.org/10.1103/PhysRevLett.93.156804

[3] C.L. Kane, E.J. Mele, Topological order and the quantum spin hall effect, Physical Review Letters 95 (2005) 146802. https://doi.org/10.1103/PhysRevLett.95.226801

[4] B. Bernevig, T. Hughes, S.C. Zhang, Quantum spin Hall effect and topological phase transition in HgTe quantum wells, Science 314 (2007) 1757-61. https://doi.org/10.1126/science.1133734

[5] M. König, Quantum spin hall insulator state in HgTe quantum wells, Science 318 (2007) 766-70. https://doi.org/10.1126/science.1148047

[6] L. Fu, C. Kane, Topological insulators in three dimensions, Physical Review Letters 98 (2007) 106803. https://doi.org/10.1103/PhysRevLett.98.106803

[7] J. Moore, L. Balents, Topological invariants of time-reversal-invariant band structures. Physical Review B 75 (2006) 121306. https://doi.org/10.1103/PhysRevB.75.121306

[8] R. Roy, Topological phases and the quantum spin Hall effect in three dimensions, Physical Review B 79 (2009) 195322. https://doi.org/10.1103/PhysRevB.79.195322

[9] Y. Ran, Y. Zhang, A. Vishwanath, One-dimensional topologically protected modes in topological insulators with lattice dislocations, Nature Physics 5 (2009) 298-303. https://doi.org/10.1038/nphys1220

[10] D. Hsieh, A topological Dirac insulator in a quantum spin Hall phase, Nature 452 (2008) 970-4. https://doi.org/10.1038/nature06843

[11] D. Hsieh, Observation of unconventional quantum spin textures in topological insulators, Science (New York, N.Y.), 2009. p. 919-22. https://doi.org/10.1126/science.1167733

[12] Y. Xia, Observation of a large-gap topological-insulator class with a single Dirac cone on the surface, Nature Physics 5 (2009) 398-402. https://doi.org/10.1038/nphys1274

[13] H. Zhang, Topological insulators in Bi2Se3, Bi2Te3 and Sb2Te3 with a single Dirac cone on the surface, Nature Physics 5 (2009) 438-442. https://doi.org/10.1038/nphys1270

[14] Y.L. Chen, Experimental Realization of a Three-Dimensional Topological Insulator, Bi2Te3, Science (New York, N.Y.), 2009. p. 178-81.

[15] A.C. Neto, The electronic properties of graphene, Review of Modern Physics 81 (2009) 109. https://doi.org/10.1103/RevModPhys.81.109

[16] P. Roushan, Topological surface states protected From backscattering by chiral spin texture, Nature 460 (2009) 1106-9. https://doi.org/10.1038/nature08308

[17] Z. Alpichshev, STM imaging of electronic waves on the surface of Bi2Te3: Topologically protected surface states and hexagonal warping effects, Physical Review Letters 104 (2010) 016401. https://doi.org/10.1103/PhysRevLett.104.016401

[18] T. Zhang, Experimental demonstration of topological surface states protected by time-reversal symmetry, Physical Review Letters 103 (2009) 266803. https://doi.org/10.1103/PhysRevLett.103.266803

[19] K. Nomura, M. Koshino, S. Ryu, Topological delocalization of two-dimensional massless Dirac Fermions, Physical Review Letters 99 (2007) 146806. https://doi.org/10.1103/PhysRevLett.99.146806

[20] P. Anderson, Absence of diffusion in certain random lattices, Physics Review 109 (1958) 1492-1505. https://doi.org/10.1103/PhysRev.109.1492

[21] J. Li, Topological Anderson insulator, Physical Review Letters 102 (2002) 136806. https://doi.org/10.1103/PhysRevLett.102.136806

[22] C. Groth, Theory of the topological Anderson insulator, Physical Review Letters 103 (2009) 196805. https://doi.org/10.1103/PhysRevLett.103.196805

[23] D. Hsieh, A tunable topological insulator in the spin helical Dirac transport regime. Nature 460 (2009) 1101-5. https://doi.org/10.1038/nature08234

[24] B. Seradjeh, J. Moore, M. Franz, Exciton condensation and charge fractionalization in a topological insulator film, Physical Review Letters 103 (2009) 066402. https://doi.org/10.1103/PhysRevLett.103.066402

[25] H. Peng, Aharonov-Bohm interference in topological insulator nanribbons. Nature Materials 9 (2009) 225-9. https://doi.org/10.1038/nmat2609

[26] Y. Zhang, Crossover of three-dimensional topological insulator of Bi2Se3 to the two-dimensional limit, Nature Physics 6 (2009) 584-588. https://doi.org/10.1038/nphys1689

[27] I. Garate, M. Franz, Inverse Spin-Galvanic Effect in a Topological-Insulator/Ferromagnet Interface. 2009.

[28] F. Wilczek, Two applications of Axon electrodynamics, Physical Review Letters 58 (1987) 1799-1802. https://doi.org/10.1103/PhysRevLett.58.1799

[29] X.L. Qi, T. Hughes, S.C. Zhang, Topological field theory of time-reversal invariant insulators, Physical Review B 81 (2008) 159901.

[30] X.L. Qi, Inducing a magnetic monopole with topological surface states, Science (New York, N.Y.) 323 (2009) 1184-7. https://doi.org/10.1126/science.1167747

[31] A. Essin, J. Moore, D. Vanderbilt, Magnetoelectric polarizability and Axion electrodynamics in crystalline insulators, Physical Review Letters 102 (2009) 146805. https://doi.org/10.1103/PhysRevLett.102.146805

[32] M. Fiebig, Multiferroics: Progress and prospects. Journal Physics D: Applied Physics (2005) 38. https://doi.org/10.1088/0022-3727/38/8/R01

[33] L. Fu, C. Kane, Superconducting proximity effect and Majorana Fermions at the surface of a topological insulator. Physical Review Letters 100 (2008) 096407. https://doi.org/10.1103/PhysRevLett.100.096407

[34] R. Jackiw, P. Rossi, Zero modes of the vortex-fermion system, Nuclear Physics B 190 (1981) 681-691. https://doi.org/10.1016/0550-3213(81)90044-4

[35] F. Wilczek, Majorana returns, Nature Physics 5 (2009) 614-618. https://doi.org/10.1038/nphys1380

Topological Insulators: Materials and Applications   Materials Research Forum LLC
Materials Research Foundations 154 (2024) 47-60   https://doi.org/10.21741/9781644902851-3

[36] J. Nilsson, A.R. Akhmerov, C. Beenakker, Splitting of a Cooper pair by a pair of Majorana bound states. Physical Review Letters 101 (2008) 120403. https://doi.org/10.1103/PhysRevLett.101.120403

[37] G. Collins, Computing with quantum knots, Scientific American 294 (2006) 56-63. https://doi.org/10.1038/scientificamerican0406-56

Materials Research Forum LLC
https://doi.org/10 21741/9781644902851-4

Chapter 4

Magnetic Topological Insulator

M. Rizwan[1*], H. Hameed[1], A. Ayub[2], H.M. Naeem Ullah[3]

[1]School of Physical Sciences, University of the Punjab, Lahore, Pakistan

[2]Department of Physics, University of Gujrat, Gujrat, Pakistan

[3]School of Materials Science and Engineering, Beijing Institute of Technology, Beijing, P. R. China

*rizwan.sps@pu.edu.pk

Abstract

In condensed matter physics, global band topology and its importance have been recognized unambiguously such as the discovery of topological insulators (TIs). Massless dispersion having spin momentum locking has been possessed by 3D topological insulators because of their bulk band topology at the surface. In Dirac band dispersion the exchange gap formation is caused by broken TRS or by the beginning of spontaneous magnetization, even though the time-reversal invariable system is the origin of 3D TIs. In such magnetic TIs, at zero magnetic fields in the exchange gap, the appearance of quantum hall effect (QHE) is the result of Fermi level tuning, and QHE at zero magnetic fields is the quantum anomalous hall effect (QAHE). In this chapter, the experimental realization and basic concepts of magnetic TIs have been discussed. The origin of magnetization in topological insulators is the main idea, which leads to QHE, and QAHE and different materials have also been discussed.

Keywords

Antiferromagnetic Phase, Quantum Anomalous Hall Effect, Topological Insulators, Ferromagnetic Phase, Intrinsic Magnetic Insulators, Integer Quantum Hall Effect (QHE)

Contents

1.   Introduction

Advanced studies have taken place in the dynamics of magnetic solids which involves conduction electrons, particularly in the quantum spin transport as this has attracted much attention in recent years, as such with the context of diluted magnetic semiconductors [1], giant magnetoresistance systems [2, 3], heavy fermion systems, colossal magnetoresistance oxides [4] as well as high-temperature superconductors [3]. Particularly industrial applications have been found in a system exhibiting colossal and giant magnetoresistance. Another branch has emerged in recent years that work on quantum transport whose concept is based on topology. Collective research has emerged in both concepts to produce new field and direction, magnetic topological insulators is an example of such a study.

Topological quantum materials have gained interest as magnetic and electronic states exhibited by such materials and integer topological invariants are used to characterize them, for example in momentum space topological electronic structures are characterized by Z_2 invariants and Chern numbers [5, 6], and in real space skyrmion numbers which are defined by spin configuration winding numbers [7]. Electronic or magnetic states of topological materials are strong under the response of external perturbation is a distinguishing feature of these materials and this is possible because of topological protection supported by the fact that continuous deformation does not affect integer topological numbers.

Integer quantum hall effect (QHE) in condensed matter in which a topological state has been observed is the main example. QHE correspondent to the conventional Hall Effect where in-hall conductance remains quantized in which a 2-dimensional electron system has been used to measure in a magnetic field [8]. TKNN (Thouless-Kohmoto- Nightingale–den Nijs) theory interrupted the experimental results with the concept of topology [9, 10]. Chern number (C) or TKNN numbers are the topological invariants that characterize broken time-reversal symmetry along with a formula for 2D electron system based on TKNN theory, in which the tendency of landau levels has corresponded to integer QHE. Hall conductivity is $\sigma_{xy} = \frac{C e^2}{h}$ where h is planks constant and e is the electron charge. As C represents an integer so no distortion of material parameters has taken place that could make a connection unceasingly to vacuum C=0 or the trivial insulators (C=0). Bulk boundary correspondence is the phenomenon responsible for the results to appear as the gapless state at the edge of a sample which formulates a number |C| that is a chiral edge channel. These channels transport the charge which is non-dissipative and Charge's sign (electron/hole) uniquely determines its direction whether it's clockwise/anticlockwise and by the applied magnetic field direction (up/down).

As per the correlation between the regular HE (brought by external magnetic field) and AHE (brought by spontaneous magnetization) an obvious question is it possible to achieve QAHE which is integer QHE at B = 0. QAHE was predicted theoretically to appear in kagome and honeycomb lattices initially with flabbergasted magnetic fluxes [11, 12]. Quantum hall insulators and gapped states are formed by these fluxes, giving conduction electron's Berry curvatures.

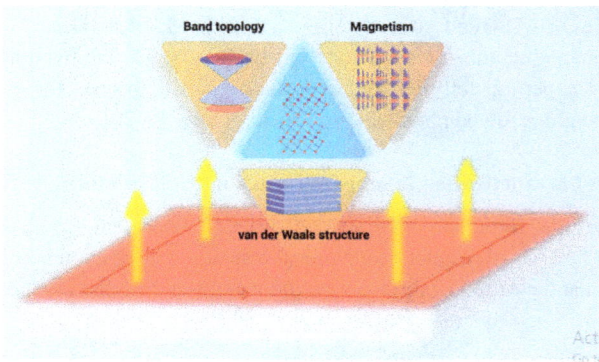

Figure 1. Magnetic topology insulators; A graphical representation [13]

Topological insulators were discovered theoretically by formulism extension for 2-D TKNN topological phase for 2 or 3 dimensional systems having broken TRS [12, 14]. Z_2 index is the topological number in this event having values $v_2=0$ 0r 1 irrespective of the TKNN number that takes any integer value. $v_2=0$ has correspondence to trivial insulators or vacuum, and $v_2=1$ has correspondence to topological insulators. Band inversion presence is the necessary condition for the advent of edge states protected topologically between the valance and conduction band by the interaction of spin-orbit which leads to band gap opening, instead of a semimetal state having overlapped bands [15]. The appearance of a 2D state at the interface having a trivial insulator (Fig. 1) or 3D topological insulator surface reflected the discontinuity of v_2 at the topological material surface. The metallic conduction band is shown by the topological insulator at the surface/interface if EF lies in between the bulk band gap but shows insulating properties in the bulk. The Hamiltonian that displays the 2d state is given by Eq.1 [16];

$$H = v_F(-k_y\sigma_x + k_x\sigma_y) \tag{1}$$

Where σ_x and σ_y are the spin Pauli matrices and v_F is the linear dispersion Fermi velocity. Massless Dirac electron's spin momentum locking is the implication of this Hamiltonian which states that an electron with opposite spin travels in contrasting directions.

Recently many 3D topological insulators have been discovered including Bi_xSb_{1-x}, as such Bi_2Sb_3, Sb_2Te_3, Bi_2Te_3, and stressed HgTe [17-19]. Massless Dirac dispersion having locking of spin momentum for such materials has been predicted hypothetically and later on confirmed experimentally as having surface-sensitive probes for example scanning tunneling or angle-resolved photoemission spectroscopy [6, 20].

2. Origin of magnetization in magnetic topological insulators

Exchange gap (EG) formation in the state of surface is important for the appearance of the developing properties of magnetic topological insulators. Interaction between spontaneous magnetization and electron in surface state induces a gap described in the mass term Eq. 2 [16]:

$$m\sigma_z = -Jn_sS_z\sigma_z \tag{2}$$

Where n_s is the localized spin areal density having average z component S_z, J is the exchange coupling among z component, unit vector is z that is normal to the surface, σ_z is the spin of Dirac electron, and S is localized spin. Magnetic interactions can be created with the surface state by an efficient approach that is chemical doping along with 3D elements of transition metal, proposed originally by theoretical study [21-23]. The study of diluted magnetic semiconductors has given a strong pillar to the doping of transition metals in topological insulators [24]. Two possible mechanisms give origin to the ferromagnetism in magnetic topological insulators: Van Vleck or local valence electron refereed Bloembergen-Rowland mechanism and carried refereed Ruderman–Kittel–Kasuya–Yosida (RKKY) mechanism [25, 26].

3.   Intrinsic magnetic TIs

There are three diverse ways to achieve magnetic TIs, the approximation to antiferromagnetic (AFM) or ferromagnetic (FM) of topological insulators, magnetic doping in topological insulators [27, 28], and the creation of ferromagnetic and antiferromagnetic orders in topological insulators [29]. New topological quantum states-QAHI (quantum anomalous hall insulators) can be generated by spontaneous magnetization produced in MTIs that unties the gap ta the Dirac point surface and interacts with the topological surface state, this happens by the tuning chemical potential up to suitable value in the thin film [30]. Dissipation of less current can be supported by the spin-polarized chiral state featured by QAHI which is very promising in energy-saving electronic applications. The first realization of QAHI was done with Cr and V-doped $(Bi,Sb)_2Te_3$ thin TI films [27, 28]. Exotic new quantum states can also be generated by the interaction among non-trivial band topology and magnetism in the magnetic topological semimetals. TRS breaking Weyl semimetal state is a great example in which spin split, linearly dispersed band cross at distinct momentum point, which results in Weyl nodes. Chiral Weyl fermions are the result of low-energy excitations close to Weyl nodes. Weyl nodes appears in pairs having conflicting chirality, and in momentum space, they can be taken as Berry curvature's source and drain. Net Berry curvature could also be represented because of broken TRS, when Weyl nodes are close to or at the Fermi level, as a result of different unusual quantum phenomena as such the ANE (Anomalous Nernst effect) [31] and intrinsic anomalous Hall effect [32]. For example magnetic and non-magnetic Weyl semi-metals can be characterized through topological surface states such as surface Fermi arcs [33]. It has been foretold by the theory that when 3D is reduced to 2D time-reversal symmetry-breaking Weyl semimetals can change to quantum anomalous Hall effect insulators [34]. Several materials have been reported experimentally including Co_2MnAl, $Co_3Sn_2S_2$, Co_2MnGa, $YbMnBi_2$, Mn_3Sn, and $GdPtBi$, to time-reversal symmetry-breaking Weyl semimetals state [29].

Topological Insulators: Materials and Applications   Materials Research Forum LLC
Materials Research Foundations 154 (2024) 61-81   https://doi.org/10.21741/9781644902851-4

3.1   Anti-ferromagnetic phase

Corresponding time-reversal symmetry (TRS) $S = \mathcal{T}t$ is sustained in magnetic configuration at ground level; here t represents half magnetic lattice conversion beside the z-axis such that nearest neighbor SLs connects with a conflicting magnetic moment. The presence of corresponding TRS represents that classification of antiferromagnetic-z $MnBi_2Te_4$ could be done into Z_2 topology. It has been shown by band structures that are calculated and associated topological invariants that bulk antiferromagnetic-z MBT ($MnBi_2Te_4$) is an antiferromagnetic TIs [35]. Topologically non-trivial energy gaps have been opened by the bulk band after spin-orbit coupling consideration. Below Neel temperature ($\sim 25K$) exchange energy gaps have been open at top surface states [36], irrespective of side surface states remaining gapless due to out-of-plane net magnetic moment [Fig. (2c)].

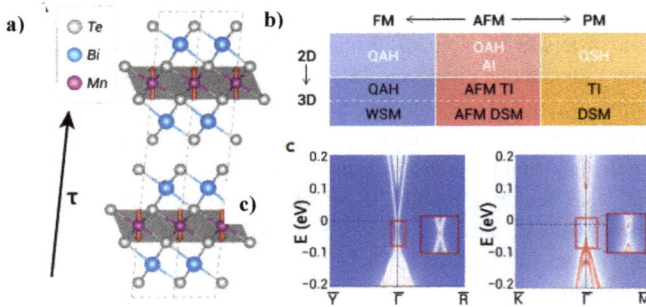

Figure 2. Calculated properties and crystal structure of $MnBi_2Te_4$ [13]

Characteristics of interlayer antiferromagnetic ordering have been retained by few-SL MBT ($MnBi_2Te_4$) [Fig. (3)]. Hence with the number of layers the topological and equivalent magnetic properties of few-SL MBT ($MnBi_2Te_4$) change. In MBT energy difference between the ferromagnetic phase and the antiferromagnetic phase that is calculated in Eq.3,

$$\Delta E_{A/_F} = E_{AFM} - E_{FM} \tag{3}$$

It is indicated by the positive $\Delta E_{A/_F}$ of monolayer ($MnBi_2Te_4$) MBT, that the ferromagnetic phase is the low energy state, hence monolayer ($MnBi_2Te_4$) MBT is a 2D ferromagnetic material [37]. Moreover, monolayer (MnBi$_2$Te$_4$) MBT is trivial topologically and has been indicated by calculated band structures [Fig. (3a)]. Even layer (MnBi$_2$Te$_4$) MBT is completely wedge AFM, in which each layer's magnetic moment has been compensated completely due to its antiferromagnetic nature.

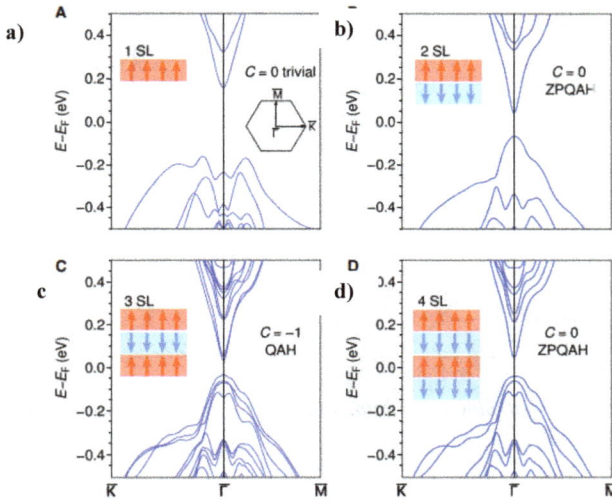

Figure 3. Designed band structure of few SL antiferromagnetic MBT [38]

Although, odd layer (MnBi$_2$Te$_4$) MBT having three or more than three layers is an uncompensated AFM demonstrating net magnetization. $|C|=1$ is the Chern insulator state of the thin film of odd layer MBT [Fig. (3c)]. For even layer MBT the Chern number is zero, however, it is non-trivial topologically. Even layer MBT's lower and upper surface give half quantum hall conductance having an opposite sign, results '0' hall conductance.

This configuration is well recognized as an axion insulator state or zero plateau QH effect [Fig. (3b, d)] [39].

3.2 Ferromagnetic phase

Transformation of antiferromagnetic MBT to Ferromagnetic ordering could happen under a reasonable magnetic field, and topological properties transformed accordingly. It is predicted that bulk material having FM-z phase is type II WSM (Weyl semimetal) phase.

These magnetic Weyl semimetals break TRS or corresponding TRS, irrespective of TR invariants Weyl semimetal having even a pair of Weyl points, and entertains only a couple of Weyl points. Hence simplest Weyl semimetal is the ferromagnetic-z MBT, which is a very beneficial upcoming experimental study of Weyl physics.

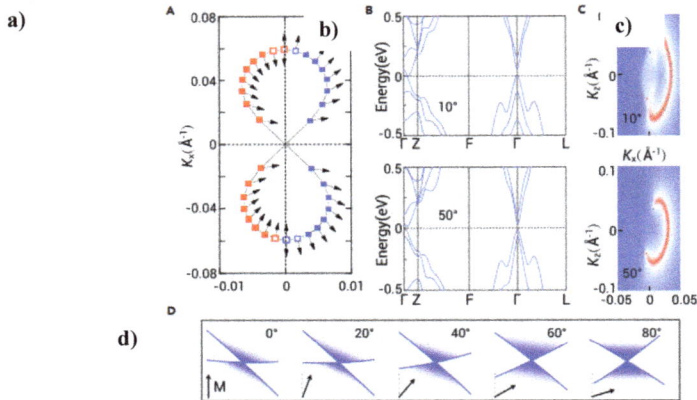

Figure 4. Under external magnetic field, topological phase transformation of bulk ferromagnetic MBT [38]

A non-trivial transport singularity which includes large anomalous Nernst effects [40], negative longitudinal magnetoresistance [41], and large intrinsic anomalous hall effects [42] is predictable in the Weyl semimetal phase. 3-fold rotational symmetry C_{3z} protects the location of pair of Weyl points that is Γ-z axis. FM-z phase of the bulk system continuously changes to the FM-x phase under the rotation of the external magnetic field, moreover, Weyl points convert to general k points deviated from the Γ-z axis in the Brillion zone due to broken C_{3z} symmetry. The system converts to type I WSM from type II WSM with polar angle θ is as such $10^0 < \theta < 20^0$ of the external magnetic field. FM-x phase turned into a trivial ferromagnetic insulator when Weyl point meets and overpower each other at $\theta = 90^0$. The Fermi arc and calculated band structure of MBT have shown in Fig.4 with $\theta = 10$ and 50 upper and lower respectively and are type I and II Weyl points. In Fig.4 Weyl points having band dispersion which is in phase direction are shown, which is a clear demonstration of transformation to type I from type II Weyl semimetals having different magnetic alignments [38].

4. Experimental observation of an intrinsic magnetic TI

QAHE has been observed in the V and Cr (Bi, Sb)$_2$Te$_3$ doped thin films, to explore the topological and fundamental physics applications the required critical temperature is below~2 K[27, 28]. For the observation of QAHE the requirement of low temperature is originated from magnetic dopants distributed randomly that induces inhomogeneous surface gaps [36]. It is predicted that high-temperature QAHE is to occur in the intrinsic antiferromagnetic and ferromagnetic TI thin film materials [43]. Until the discovery has been done of an intrinsic antiferromagnetic TI MnBi$_2$Te$_4$ [44], there was very minor progress even though a significant effort has been made in experimental and theoretical research. MnBi$_2$Te$_4$ is a composite of layered ternary tetradymite. Rhombohedral (R-3m) is the structure in which it crystallizes. Assembled as piling of Te-Bi-Te-Mn-Te-Bi-Te (SLs) (Fig. 5(a)) septuple layers. Van der Waals bonding by which SLs are coupled.

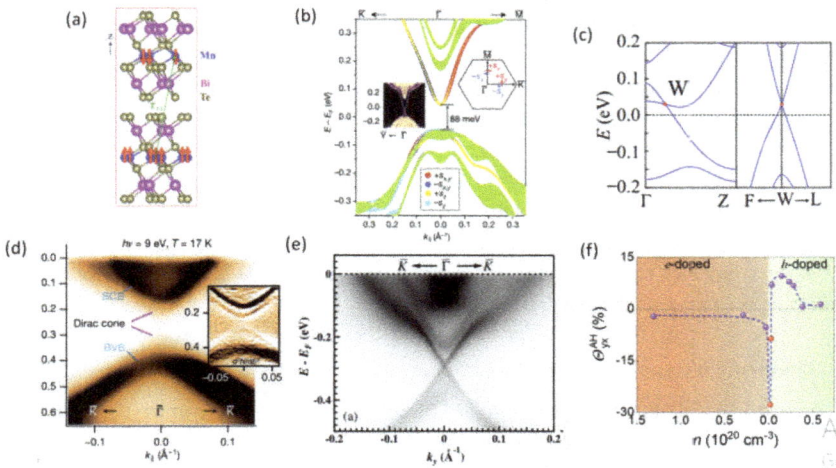

Figure 5. Magnetic and electronic structure of MnBi$_2$Te$_4$ along with band dispersion.[29]

There are two methods to grow MnBi$_2$Te$_4$; by using the flux method [43] in which excessive flux has been served by Bi$_2$Te$_4$, or from melting through stoichiometric composition [36]. Single crystal of MnBi$_2$Te$_4$ can be attained only with quenching at a temperature near 590°C as it is metastable. To achieve melt growth, it is needed to heat the stoichiometric mixture at a high temperature of almost 700°C-1000°C and needs to cool it down slowly close the temperature to 590°C, and lastly follow through with quenching and annealing at 590°C [36]. To achieve the flux growth method continued slow cooling for almost two weeks at temperatures ~600°C to ~590°C is essential, and through centrifuging unnecessary flux is separated [43]. AFM combines with non-trivial band topology assisted through $MnBi_2Te_4$ [36], which gives raise to intrinsic antiferromagnetic TI. Mn sublattice produces AFM and inverted Bi and Te p_z bands forms non-trivial band topology at Γ points because of strong SOC (spin-orbit coupling). A-type antiferromagnetic order (T_N=25K) [43] is shown by an antiferromagnetic state, characterized by Mn ferromagnetic layers loaded anti-Ferro magnetically along the c-axis, and alignment of magnetic moment is along the c-axis. Recently the experiment related to inelastic neutron scattering conducted on $MnBi_2Te_4$ explored magnetic frustration and large spin gap because of the maximum subsequent nearest neighboring antiferromagnetic exchange of $MnBi_2Te_4$ [36]. A large gap on the 001 surfaces is opened because of the breaking of the $S = \theta T_{1/2}$ at surface Dirac nodes, here θ is TRS and $T_{1/2}$ is the PTS (primitive translation symmetry) [Fig. 5(b)]. This type of gap was explored by Otrokov et al in ARPES measurement preceded by a single crystal and by other groups subsequently [Fig. 5(d)]. Though it has been shown that either in antiferromagnetic or paramagnetic state surface Dirac cone is gapless which is reported occasionally in ARPES experiments [Fig. 5(e)]. An ideal platform has been offered by $MnBi_2Te_4$ for the realization of a new exotic TQS (topological quantum state). $MnBi_2Te_4$ is not only capable of hosting in thin films samples axion insulator having topological magnetoelectric effects and high-temperature QAHE but also perfect WSM state having one pair of Weyl nodes close to Fermi level, derived by strain [Fig. 5(c)] or external magnetic field in its bulk ferromagnetic state, predicted theoretically [36]. Furthermore, it has also been predicted chiral Majorana mode is accessible through interaction among s-wave superconductor and $MnBi_2Te_4$. Remarkable progress towards the realization of these predicted QS (quantum state) has been made recently [29].

5.   Quantum anomalous hall effect in magnetic TIs

It is known in solid-state transport regular HE and AHE are the outcomes of the breaking of time-reversal symmetry. Precisely, in previous cases, there has been a requirement for the perpendicular magnetic field to generate transverse voltage V_{xy} for the detection of conduction charge particles [45], Latter on spin-orbit interactions among Magnetic moment and charge currents cases V_{xy} and external magnetic field replaced by spontaneous magnetization. Under high magnetic field observation of quantized dissipation less chiral edge conduction in 2D high mobility electron gas, since quantum hall effect discovered in 1980 [46]. On the other hand, after Haldane proposed 1st theoretical model, for the realization of QAHE very little progress was made [8]. Research of quantum anomalous hall effect and appropriate non-zero Chern insulators was made practical until TI materials were discovered.

5.1 Quantum spin hall effect in 2D system

The motivation for the construction of QAHE is the result of quantum spin hall effect (QSHE)realization in the HgTe/CdTe 2-Dimension TI system [47]. As these two phenomena are closely related, it is claimed in the QSHE regime helical edge state could be observed as 2 duplicates of the QAHE having opposite chirality. A single quantum anomalous HE is observed if one spin block is repressed because of time reversal symmetry breaking. Introduction of extra Mn ions into the quantum wells of HgTe There has been a change in the topological surface band, built on 2d TI four-band models. Here spin splitting term can be expanded induced by magnetization into phenomenological form given as in Eq. 4 [48]:

$$H_S = \begin{bmatrix} G_E & 0 & 0 & 0 \\ 0 & G_H & 0 & 0 \\ 0 & 0 & -G_E & 0 \\ 0 & 0 & 0 & -G_H \end{bmatrix} \tag{4}$$

For two electron sub-bands $|E_1, \pm >$, a difference of split energy is $2G_E$ and for two hole sub-bands $|H_1, \pm >$, it is $2G_H$, $|E_1, \pm>$ and $|H_1, \pm>$ will have an opposite sign for the condition $G_E \times G_H < 0$, such that suppose $G_E < 0$ and $G_H > 0$ having large splitting energy, the band transportation of $|E_1, ->$, and $|H_1, ->$ states, will finally arrive in the normal region and quantum transport region only spin-up edge states are left while $|H_1, + >$ and $|E_1, + >$ still holds [Fig. 6(a)]. Later it was pointed out by Yu et al by following a similar spin-splitting situation (i.e; $G_E \times G_H > 0$), actually 4 band system was in a topologically trivial phase, spin-down sub-bands $|E_1, ->$, and $|H_1, ->$, are pushed away further from one another when band inversion can be induced in spin up sub-bands $|H_1, + >$ and $|E_1, + >$ with appropriate exchange field [Fig. 6(b)] [49].

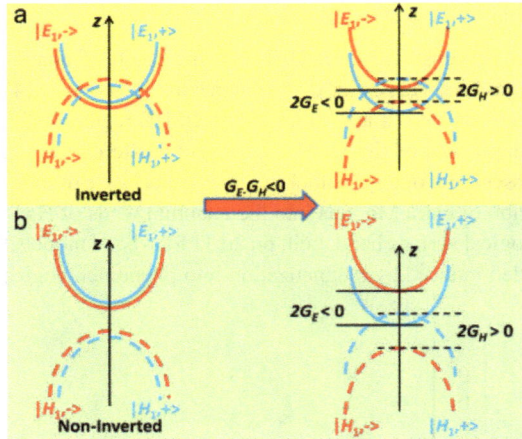

Figure 6. Inverted and non-inverted band evolution upon increment in spin splitting
[50]

Construction of a single quantum anomalous hall effect is possible under such conditions. Necessary spin splitting energy failed to achieve because spontaneously $Hg_{1-x}Mn_xTe$ was not ordered by Mn moments even though Liu et al. model was practicable. On the other hand, bulk Van Vleck susceptibility generates definite magnetization in the TI Cr-doped films nevertheless if majority achieves an insulating state, in the Cr-doped tetradymite-type topological insulating system, for the realization of QAHE improved model was proposed by Yu et al. [49]. It was found that quantum tunneling among the bottom and top surfaces is possible when the thickness of Cr-doped 3-dimensional regimes is reduced to 2D hybridization that rose the finite mass term: $m_k = m_0 + B(k_x^2 + k_y^2)$ and modification of surface Hamiltonian are possible as given in Eq. 5 [49]:

$$H = \begin{bmatrix} h_k + gM\sigma_z & 0 \\ 0 & h_k^* - gM\sigma_z \end{bmatrix} \tag{5}$$

Where $h_k = m_k\sigma_z + v_F(k_y\sigma_x - k_x\sigma_y)$, v_F, M, and g have the same values for both bottom and top surfaces under the assumption of spatial inversion symmetry, hence $G_E \times G_H < 0$ was achieved automatically [51].

5.2   QHE, QSHE, and QAHE

As it is clear that the intrinsic mechanism of QHE, QSHE, and QAHE is quite distinct although they all fit the quantized edge transport phenomenon. In this section comparison of the relation and differences among quantum trio and summarization has been discussed, and for low-power interrelated applications the advantage of QAHE has been shown. It is well known that the realization of QHE can be held on $C_1=1$ Chern insulator, in which due to discrete Landau levels bulk is insulating and conduction channels exist at the edge only. The external magnetic field in the QHE regime acts as a gauge field that couples with electron moment and in real space urge the electron to create cyclotron motions. Cyclotron motion circles of an electron need to be completed to build discrete LLs before the momentum of the electron is lost because of scattering, this can only be achieved when the electron has very high mobility [Fig.7(a)]. It is needed for the electron to be confined in 2D for précised LL quantization so that energy levels can take discrete values $E_n = \hbar\omega_c(1 + \frac{1}{2})$ where $\omega_c = \frac{eB}{m}$ is cyclotron frequency. The chiral edge states appear eagerly when among two neighboring LLs Fermi level is located while through cyclotron orbitals bulk carriers are localized. More specifically, high mobility material having good confinement of 2D and strong magnetization is required to realize QHE [52].

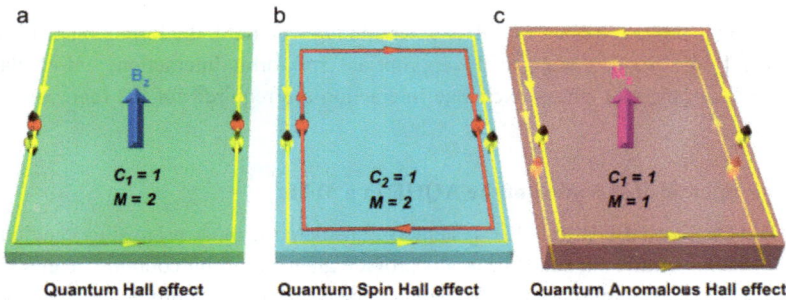

Figure 7. QHE, QSHE, and QAHE [53]

Conversely, an external magnetic field doesn't require for QSHE. It is observed in systems where band inversion at Γ points happen through strong SOC in BZ. The system becomes a chiral insulator ($C_2=1$) through this band inversion, having M=2 quantized channel number. Here topology is described through the second Chern number in the QSHE state [Fig. 7(b)]. The reason behind this is that in such systems the TRS is preserved, and the subsequent instead of chiral quantum edge state is helical. Quantized $2e^2/h$ helical conduction is exhibited by the longitudinal conductance once the Fermi level is positioned in the band gap of bulk. Since spin-orbit interaction strength determines the bulk band gap is typically minor (i.e.; up to 0.3 eV), to band potential fluctuations and carrier states induces by intrinsic bulk impurities, quantum edge helical conduction is very small. Therefore, high material quality is required by QSHE (i.e.; high mobility and low bulk carrier density. Till now experimentally QSHE has been detected only in InAs/GaSb and HeTe/CdTe QWs (quantum well) [54, 55].

In comparison with QHE and QSHE, both magnetic exchange interactions and spin-orbit interaction involves in QAHE in a system of magnetic insulation. In the region of QAHE having a forceful magnetic moment, band inversion is the result of strong spin-orbit interconnection. However, from the impulsive magnetization large exchange field (i.e; 100 T) couples with band electron spin, and separate the spin up or down bands in different orientations. As a result, one pair remains inside the non-trivial topological state while keeping the additional pair un-inverted of spin-resolved bands. This results in the change of bulk topology in comparison to the QSHE state, and the system converts into Chern insulator $C_1=1$, and transformation at 0 external magnetic fields of quantized QAHE chiral edge can be attained if the location of Fermi level is in huge band gap of Dirac surface. In the meantime, Time reversal symmetry in QAHE in comparison with QHE by spontaneous magnetization is broken irrespective of an external magnetic field, the formation of LLs is not needed [Fig. 7(c)]. Conclusively appropriate spin-orbit interaction, out-of-plane magnetic anisotropy, and strong exchange interaction are required for the realization of QAHE[56].

6.   Experimental observation of the AQHE in a MTIs

For QAHE Cr-doped Bi_2Te_3/Sb_2Te_3 system is considered a most promising candidate, because of the inverted bulk band, Te atoms provide giant spin-orbit coupling, Van Vleck mechanism helps to establish large out-of-plane magnetization instantly without the help of wandering carrier, through gate modulation it is easy to tune Fermi level across Dirac points, and especially by optimizing the Bi/Sb ratio it is possible to minimize the carrier density practically. The first observation of QAHE was reported by Chang et al. from Tsinghua on 5QL $Cr_{0.15}(Bi_{0.1}Sb_{0.9})_{1.85}Te_3$ and the growth was taken on $SrTiO_3$ (111) substrate [Fig. 8(a,c)] [56]. They achieved QAHE at 30 mk in their work. More precisely, into the gap surface ($V_g \sim -1.5$), they electrically tuned the Fermi level, and at h/e^2 (25.8 kΩ), R_{xy} was quantized and it was suggested that charge neutrality and edge conduction of thin film was perfect because with an external magnetic field such quantized value was almost invariant [57, 58]. With the help of a highly dissipative bulk channel, they observed giant MR (2251%) and then related it to quantum phase transition among two different QAHE states [Fig. 8(c)]. Moreover, at zero filed gate dependent R_{xx} and R_{xy} were provided [Fig. 8(b)], distinct plateau exhibited by R_{xy} having quantized value h/e^2, and sharp dip down was experienced by longitudinal R_{xy} to 0.098 h/e^2 correspondingly.

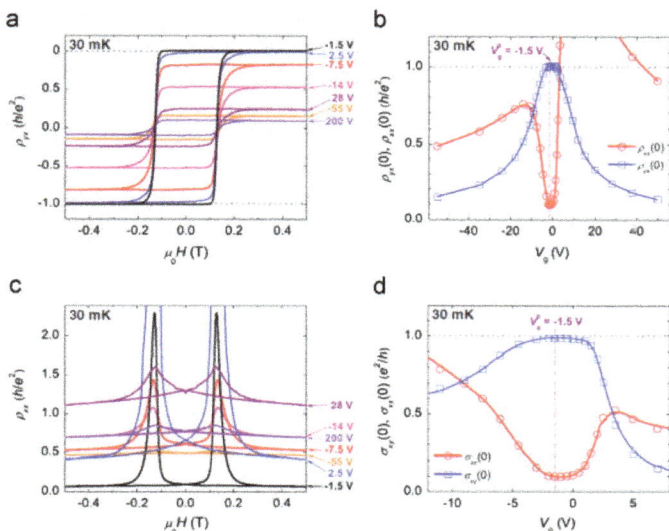

Figure 8. Experimental observation of the QAHE in a MTIs [53]

These transport data were fitted with a real conduction model by Lu et al. recently and confirmed with an agreement among theory and experiment that topological non-trivial conduction band originated transformation in the QAHE regime, which in group velocity had concentrated local maximum and Berry curvature. At zero field in the QAHE regime, non-zero longitudinal resistance R_{xx} was detected through Chang et al. experiment on the 5 QL flick. By applying magnetic field B>10T (large and perpendicular) further into the perfect QHE region 5QL film was derived and R_{xx} vanished very close to zero[56]. They further analyzed the roots of such phenomena, both gapless quasi-helical edge states and variable range hopping (VRH) have been proposed in the 5QL Cr- doped TI film to clarify such field-dependent R_{xx} outcomes.

Conclusion

Lately, there has been progress in the area of topological science through groundbreaking theoretical findings, which inspired the experimental studies of interesting phenomena in TI. In this chapter, the concept along with the experimental realization of the magnetic TIs has been discussed thoroughly. Massive Dirac gap and QAHS's materialization along with axion insulator state has been realized. From the perspective of material science, magnetic doping of TI is the main subject of the massive Dirac gap with broken TRS to induce long-range ferromagnetic order in material. In the Dirac surface state, interactions among local magnetic moment and surface state electrons in Cr-doped $(Bi, Sb)_2Te_3$ open exchange gap.

References

[1] Y.J. Tokura, Critical features of colossal magnetoresistive manganites, Reports on Progress in Physics 69 (2006) 797. https://doi.org/10.1088/0034-4885/69/3/R06

[2] S.G. Stewart, Heavy-fermion systems, Reviews of Modern Physics 56 (1984) 755. https://doi.org/10.1103/RevModPhys.56.755

[3] P.A. Lee, N. Nagaosa, X.G. Wen, Doping a Mott insulator: Physics of high-temperature superconductivity, Reviews of Modern Physics 78 (2006) 17-85. https://doi.org/10.1103/RevModPhys.78.17

[4] A. Fert, Nobel lecture: Origin, development, and future of spintronics, Reviews of Modern Physics 80 (2008) 1517-1530. https://doi.org/10.1103/RevModPhys.80.1517

[5] M.Z. Hasan, C.L. Kane, Colloquium: Topological insulators, Reviews of Modern Physics 82 (2010) 3045-3067. https://doi.org/10.1103/RevModPhys.82.3045

[6] X.L. Qi, Zhang, Topological Insulators and superconductors 83 (2011) 1057. https://doi.org/10.1103/RevModPhys.83.1057

[7] N. Nagaosa, Y.J.N.n. Tokura, Topological properties and dynamics of magnetic skyrmions, Nature Nanotechnology 8 (2013) 899-911. https://doi.org/10.1038/nnano.2013.243

[8] K.V. Klitzing, G. Dorda, M. Pepper, New Method for High-Accuracy Determination of the Fine-Structure Constant Based on Quantized Hall Resistance, Physical Review Letters 45 (1980) 494-497. https://doi.org/10.1103/PhysRevLett.45.494

[9] D.J. Thouless, M. Kohmoto, M.P. Nightingale, M. den Nijs, Quantized Hall Conductance in a Two-Dimensional Periodic Potential, Physical Review Letters 49 (1982) 405-408. https://doi.org/10.1103/PhysRevLett.49.405

[10] M. Kohmoto, Topological invariant and the quantization of the Hall conductance, Annals of Physics 160 (1985) 343-354. https://doi.org/10.1016/0003-4916(85)90148-4

[11] C.-T. Ho, D.-W. Wang, Robust identification of topological phase transition by self-supervised machine learning approach, New Journal of Physics 23 (2021) 083021. https://doi.org/10.1088/1367-2630/ac1709

[12] K. Ohgushi, S. Murakami, N. Nagaosa, Spin anisotropy and quantum Hall effect in the kagom\'e lattice: Chiral spin state based on a ferromagnet, Physical Review B 62 (2000) R6065-R6068. https://doi.org/10.1103/PhysRevB.62.R6065

[13] P. Wang, J. Ge, J. Li, Y. Liu, Y. Xu, J.J.T.I. Wang, Intrinsic magnetic topological insulators, 2 (2021) 100098. https://doi.org/10.1016/j.xinn.2021.100098

[14] C.L. Kane, E.J.J.P.r.l. Mele, Quantum spin Hall effect in graphene, 95 (2005) 226801. https://doi.org/10.1103/PhysRevLett.95.226801

[15] Y. Tokura, K. Yasuda, A. Tsukazaki, Magnetic topological insulators, Nature Reviews Physics 1 (2019) 126-143. https://doi.org/10.1038/s42254-018-0011-5

[16] Y. Tokura, K. Yasuda, A.J.N.R.P. Tsukazaki, Magnetic topological insulators, 1 (2019) 126-143. https://doi.org/10.1038/s42254-018-0011-5

[17] Y.J.J.o.t.P.S.o.J. Ando, Topological insulator materials, 82 (2013) 102001. https://doi.org/10.7566/JPSJ.82.102001

[18] R.J. Cava, H. Ji, M.K. Fuccillo, Q.D. Gibson, Y.S. Hor, Crystal structure and chemistry of topological insulators, Journal of Materials Chemistry C 1 (2013) 3176-3189. https://doi.org/10.1039/c3tc30186a

[19] H. Zhang, C.-X. Liu, X.-L. Qi, X. Dai, Z. Fang, S.-C. Zhang, Topological insulators in Bi2Se3, Bi2Te3 and Sb2Te3 with a single Dirac cone on the surface, Nature Physics 5 (2009) 438-442. https://doi.org/10.1038/nphys1270

[20] M.Z. Hasan, C.L.J.R.o.m.p. Kane, Colloquium: topological insulators, 82 (2010) 3045. https://doi.org/10.1103/RevModPhys.82.3045

[21] R. Yu, W. Zhang, H.J. Zhang, S.C. Zhang, X. Dai, Z. Fang, Quantized anomalous Hall effect in magnetic topological insulators, Science (New York, N.Y.) 329 (2010) 61-64. https://doi.org/10.1126/science.1187485

[22] R.R. Biswas, A.V. Balatsky, Impurity-induced states on the surface of three-dimensional topological insulators, Physical Review B 81 (2010) 233405. https://doi.org/10.1103/PhysRevB.81.233405

[23] G. Rosenberg, M. Franz, Surface magnetic ordering in topological insulators with bulk magnetic dopants, Physical Review B 85 (2012) 195119. https://doi.org/10.1103/PhysRevB.85.195119

[24] J.-M. Zhang, W. Zhu, Y. Zhang, D. Xiao, Y. Yao, Tailoring Magnetic Doping in the Topological Insulator Bi2Se3, Physical Review Letters 109 (2012) 266405. https://doi.org/10.1103/PhysRevLett.109.266405

[25] Q. Liu, C.-X. Liu, C. Xu, X.-L. Qi, S.-C. Zhang, Magnetic Impurities on the Surface of a Topological Insulator, Physical Review Letters 102 (2009) 156603. https://doi.org/10.1103/PhysRevLett.102.156603

[26] D.A. Abanin, D.A. Pesin, Ordering of Magnetic Impurities and Tunable Electronic Properties of Topological Insulators, Physical Review Letters 106 (2011) 136802. https://doi.org/10.1103/PhysRevLett.106.136802

[27] C.Z. Chang, J. Zhang, X. Feng, J. Shen, Z. Zhang, M. Guo, K. Li, Y. Ou, P. Wei, L.L. Wang, Z.Q. Ji, Y. Feng, S. Ji, X. Chen, J. Jia, X. Dai, Z. Fang, S.C. Zhang, K. He, Y. Wang, L. Lu, X.C. Ma, Q.K. Xue, Experimental observation of the quantum anomalous Hall effect in a magnetic topological insulator, Science (New York, N.Y.) 340 (2013) 167-170. https://doi.org/10.1126/science.1234414

[28] C.Z. Chang, W. Zhao, D.Y. Kim, H. Zhang, B.A. Assaf, D. Heiman, S.C. Zhang, C. Liu, M.H. Chan, J.S. Moodera, High-precision realization of robust quantum anomalous Hall state in a hard ferromagnetic topological insulator, Nature materials 14 (2015) 473-477. https://doi.org/10.1038/nmat4204

[29] W. Ning, Z.J.A.M. Mao, Recent advancements in the study of intrinsic magnetic topological insulators and magnetic Weyl semimetals, 8 (2020) 090701. https://doi.org/10.1063/5.0015328

[30] P. Wei, F. Katmis, B.A. Assaf, H. Steinberg, P. Jarillo-Herrero, D. Heiman, J.S. Moodera, Exchange-Coupling-Induced Symmetry Breaking in Topological Insulators, Physical Review Letters 110 (2013) 186807. https://doi.org/10.1103/PhysRevLett.110.186807

[31] H.K. Singh, I. Samathrakis, N.M. Fortunato, J. Zemen, C. Shen, O. Gutfleisch, H. Zhang, Multifunctional antiperovskites driven by strong magnetostructural coupling, npj Computational Materials 7 (2021) 98. https://doi.org/10.1038/s41524-021-00566-w

[32] A.J.P.r.l. Burkov, Anomalous Hall effect in Weyl metals, 113 (2014) 187202. https://doi.org/10.1103/PhysRevLett.113.187202

[33] X. Wan, A.M. Turner, A. Vishwanath, S.Y.J.P.R.B. Savrasov, Topological semimetal and Fermi-arc surface states in the electronic structure of pyrochlore iridates, 83 (2011) 205101. https://doi.org/10.1103/PhysRevB.83.205101

[34] G. Xu, H. Weng, Z. Wang, X. Dai, Z.J.P.r.l. Fang, Chern semimetal and the quantized anomalous Hall effect in HgCr 2 Se 4, 107 (2011) 186806. https://doi.org/10.1103/PhysRevLett.107.186806

[35] M.M. Otrokov, I.P. Rusinov, M. Blanco-Rey, M. Hoffmann, A.Y. Vyazovskaya, S.V. Eremeev, A. Ernst, P.M. Echenique, A. Arnau, E.V. Chulkov, Unique Thickness-Dependent Properties of the van der Waals Interlayer Antiferromagnet ${\mathrm{MnBi}}_{2}{\mathrm{Te}}_{4}$ Films, Physical Review Letters 122 (2019) 107202. https://doi.org/10.1103/PhysRevLett.122.107202

[36] M.M. Otrokov, Klimovskikh, II, H. Bentmann, D. Estyunin, A. Zeugner, Z.S. Aliev, S. Gaß, A.U.B. Wolter, A.V. Koroleva, A.M. Shikin, M. Blanco-Rey, M. Hoffmann, I.P. Rusinov, A.Y. Vyazovskaya, S.V. Eremeev, Y.M. Koroteev, V.M. Kuznetsov, F. Freyse, J. Sánchez-Barriga, I.R. Amiraslanov, M.B. Babanly, N.T. Mamedov, N.A. Abdullayev, V.N. Zverev, A. Alfonsov, V. Kataev, B. Büchner, E.F. Schwier, S. Kumar, A. Kimura, L. Petaccia, G. Di Santo, R.C. Vidal, S. Schatz, K. Kißner, M. Ünzelmann, C.H. Min, S. Moser, T.R.F. Peixoto, F. Reinert, A. Ernst, P.M. Echenique, A. Isaeva, E.V. Chulkov, Prediction and observation of an antiferromagnetic topological insulator, Nature 576 (2019) 416-422. https://doi.org/10.1038/s41586-019-1840-9

[37] K.S. Burch, D. Mandrus, J.G. Park, Magnetism in two-dimensional van der Waals materials, Nature 563 (2018) 47-52. https://doi.org/10.1038/s41586-018-0631-z

[38] P. Wang, J. Ge, J. Li, Y. Liu, Y. Xu, J. Wang, Intrinsic magnetic topological insulators, The Innovation 2 (2021) 100098. https://doi.org/10.1016/j.xinn.2021.100098

[39] D. Zhang, M. Shi, T. Zhu, D. Xing, H. Zhang, J. Wang, Topological Axion States in the Magnetic Insulator MnBi2Te4 with the Quantized Magnetoelectric Effect, Physical Review Letters 122 (2019) 206401. https://doi.org/10.1103/PhysRevLett.122.206401

[40] L. Ding, J. Koo, L. Xu, X. Li, X. Lu, L. Zhao, Q. Wang, Q. Yin, H Lei, B. Yan, Z. Zhu, K. Behnia, Intrinsic Anomalous Nernst Effect Amplified by Disorder in a Half-Metallic Semimetal, Physical Review X 9 (2019) 041061. https://doi.org/10.1103/PhysRevX.9.041061

[41] E. Liu, Y. Sun, N. Kumar, L. Müchler, A. Sun, L. Jiao, S.Y. Yang, D. Liu, A. Liang, Q. Xu, J. Kroder, V. Süß, H. Borrmann, C. Shekhar, Z. Wang, C. Xi, W. Wang, W.

Schnelle, S. Wirth, Y. Chen, S.T.B. Goennenwein, C. Felser, Giant anomalous Hall effect in a ferromagnetic Kagomé-lattice semimetal, Nat Phys 14 (2018) 1125-1131. https://doi.org/10.1038/s41567-018-0234-5

[42] Q. Wang, Y. Xu, R. Lou, Z. Liu, M. Li, Y. Huang, D. Shen, H. Weng, S. Wang, H. Lei, Large intrinsic anomalous Hall effect in half-metallic ferromagnet Co(3)Sn(2)S(2) with magnetic Weyl fermions, Nature communications 9 (2018) 3681. https://doi.org/10.1038/s41467-018-06088-2

[43] J.Q. Yan, Q. Zhang, T. Heitmann, Z. Huang, K.Y. Chen, J.G. Cheng, W. Wu, D. Vaknin, B.C. Sales, R.J. McQueeney, Crystal growth and magnetic structure of MnBi2Te4, Physical Review Materials 3 (2019) 064202. https://doi.org/10.1103/PhysRevMaterials.3.064202

[44] J. Li, Y. Li, S. Du, Z. Wang, B.L. Gu, S.C. Zhang, K. He, W. Duan, Y. Xu, Intrinsic magnetic topological insulators in van der Waals layered MnBi(2)Te(4)-family materials, Science advances 5 (2019) eaaw5685. https://doi.org/10.1126/sciadv.aaw5685

[45] E.H. Hall, XVIII. On the "Rotational Coefficient" in nickel and cobalt, The London, Edinburgh, and Dublin Philosophical Magazine and Journal of Science 12 (1881) 157-172. https://doi.org/10.1080/14786448108627086

[46] F.D.M.J.P.r.l. Haldane, Model for a quantum Hall effect without Landau levels: Condensed-matter realization of the" parity anomaly", 61 (1988) 2015. https://doi.org/10.1103/PhysRevLett.61.2015

[47] M. König, S. Wiedmann, C. Brüne, A. Roth, H. Buhmann, L.W. Molenkamp, X.L. Qi, S.C. Zhang, Quantum spin hall insulator state in HgTe quantum wells, Science (New York, N.Y.) 318 (2007) 766-770. https://doi.org/10.1126/science.1148047

[48] C.-X. Liu, X.-L. Qi, X. Dai, Z. Fang, S.-C.J.P.r.l. Zhang, Quantum anomalous Hall effect in Hg 1− y Mn y Te quantum wells, 101 (2008) 146802. https://doi.org/10.1103/PhysRevLett.101.146802

[49] R. Yu, W. Zhang, H.-J. Zhang, S.-C. Zhang, X. Dai, Z.J.s. Fang, Quantized anomalous Hall effect in magnetic topological insulators, 329 (2010) 61-64. https://doi.org/10.1126/science.1187485

[50] X. Kou, Y. Fan, M. Lang, P. Upadhyaya, K.L. Wang, Magnetic topological insulators and quantum anomalous hall effect, Solid State Communications 215-216 (2015) 34-53. https://doi.org/10.1016/j.ssc.2014.10.022

[51] H.-Z. Lu, W.-Y. Shan, W. Yao, Q. Niu, S.-Q.J.P.r.B. Shen, Massive Dirac fermions and spin physics in an ultrathin film of topological insulator, 81 (2010) 115407. https://doi.org/10.1103/PhysRevB.81.115407

[52] S. Datta, Electronic transport in mesoscopic systems, Cambridge university press1997.

[53] X. Kou, Y. Fan, M. Lang, P. Upadhyaya, K.L.J.S.S.C. Wang, Magnetic topological insulators and quantum anomalous hall effect, 215 (2015) 34-53. https://doi.org/10.1016/j.ssc.2014.10.022

[54] L. Du, I. Knez, G.J. Sullivan, R.-R.J.B.o.t.A.P.S. Du, Observation of Quantum Spin Hall States in InAs/GaSb Bilayers under Broken Time-Reversal Symmetry, 2014 (2013).

[55] A. Roth, C. Brüne, H. Buhmann, L.W. Molenkamp, J. Maciejko, X.L. Qi, S.C. Zhang, Nonlocal transport in the quantum spin Hall state, Science (New York, N.Y.) 325 (2009) 294-297. https://doi.org/10.1126/science.1174736

[56] C.-Z. Chang, J. Zhang, X. Feng, J. Shen, Z. Zhang, M. Guo, K. Li, Y. Ou, P. Wei, L.-L. Wang, Z.-Q. Ji, Y. Feng, S. Ji, X. Chen, J. Jia, X. Dai, Z. Fang, S.-C. Zhang, K. He, Y. Wang, L. Lu, X.-C. Ma, Q.-K. Xue, Experimental Observation of the Quantum Anomalous Hall Effect in a Magnetic Topological Insulator, 340 (2013) 167-170. https://doi.org/10.1126/science.1234414

[57] Transport properties of topological insulators films and nanowires, Chinese Physics B 22 (2013) 067302. https://doi.org/10.1088/1674-1056/22/6/067302

[58] X. Kou, L. He, M. Lang, Y. Fan, K. Wong, Y. Jiang, T. Nie, W. Jiang, P. Upadhyaya, Z. Xing, Y. Wang, F. Xiu, R.N. Schwartz, K.L. Wang, Manipulating Surface-Related Ferromagnetism in Modulation-Doped Topological Insulators, Nano Letters 13 (2013) 4587-4593. https://doi.org/10.1021/nl4020638

Topological Insulators: Materials and Applications
Materials Research Foundations 154 (2024) 82-94

Materials Research Forum LLC
https://doi.org/10.21741/9781644902851-5

Chapter 5

Topological Superconductor

M. Rizwan[1*], H. Hameed[1], H.M. Naeem Ullah[2], A. Ayub[3]

[1]School of Physical Sciences, University of the Punjab, Lahore, Pakistan

[2]School of Materials Science and Engineering, Beijing Institute of technology, Beijing, P. R. China

[3]Department of Physics, University of Gujrat, Pakistan

*rizwan.sps@pu.edu.pk-

Abstract

In the scientific community, in topological materials superconductivity has attracted scientists significantly as Majorana fermions as observed in these materials perceived from zero-biased conduction peak, quantized thermal conductivity, and AJE (Anomalous Josephson Effect). In this chapter recent advancements in the field of TSCs have been discussed. Role Majorana fermions in TSCs has been discussed briefly. Few materials that show topological superconductivity under specific conditions have also been considered. Unconventional doping-based sed superconductors showed topological superconductivity having topological invariant states in the form of bulk from under certain conditions.

Keywords

Superconductors, Majorana Fermions, Quantum Spin, Spin Current, Nematicity, Josephson Effect

Contents

1.   Introduction

Quantum matter classifies magnets, crystalline solids, and superconductors as having exceptional states. The study of Spontaneous symmetry breaking is responsible for the classification of these states. There happens to be ordered parameters that in ordered states have non-zero exception values. Ginzburg Landau's (G-L) theory determines order parameters and uniquely describes these quantum states[1, 2]. Conversely, quantum matter in topology creates new states with ordered parameters, that do not lie in the region of Ginzburg Landau (G-L) theory. In the 1980's, such type of quantum states has been discovered by Von Klitzing and named these states quantum hall states in 2D system. Insulating gap is the center of these states, while skipping orbits makes edge conducting, which forms because of externally applied magnetic field. Single directional dissipation less path to electric current provided by these edge states that raises quantized hall effect [3]. There are certain parameters in topological phases that are unresponsive under smooth deformation in the Hamiltonian, and these parameters are topological invariants. Classification of smooth deformation in the surface states, as the change in which bulk energy remains undisturbed. Furthermore, in specific topological phases significant role is played by energy gap in bulk, which raises two new modules of quantum materials, and these are topological superconductors TSC and topological insulators TI's. In particular, conductors and doped semiconductors does not observe these effects as they do not have energy gap [4]. C.L. Kane et al. first time predicted Topological insulators theoretically in 2005 [5], and in 2007 Koing et.al realized them experimentally in HgTe/CdTe quantum wells [6]. From quantum hall system TIs are dissimilar as in presence of impurity TRS (time reversal symmetry) is broken. And important role is played by spin-orbit interaction as TRS is preserved. TIs also named QSH (quantum spin hall) systems because of intrinsic

SOC (spin-orbit coupling) presence. Quantum spin hall states are non-trivial topologically and contain conducting surface states and fully insulating bulk having Dirac fermion's odd number [7]. At least double degeneracy is important for the states that have an odd number of fermions, which states that in a topological invariant (non-trivial) system surface positions are paired. These states move in different directions to each other [8].

In fault-tolerant quantum computing topological superconductors contains greater importance due to Majorana fermions emergence [9]. The classification of topological superconductors is based on type of superconductivity and time reversal symmetry presence. In the previous on, topological superconductors are classified as weak and strong topological superconductors. Nodal superconducting gap happens to be in some exceptional superconductors. To consider topological superconductors as weak then nodal superconductors should have such topological properties [1]. Three main characterizations are shown by strong topological superconductors: 1) Conducting gapless surface states, 2) Full superconducting gap having odd parity pairing symmetry, 3) In superconducting vortex cores the Majorana Zero Modes (MZMs) [1]. In the perspective of TRS, there are two categories of topological superconductors, one is in which TRS is preserved and in other it is broken [10, 11]. In earlier type of topological superconductors, time reversal symmetry breaks due to internal magnetic field in superconductors having odd parity pairing, such as Sr_2RuO_4 [12]. While in the other class, presence of strong orbital coupling preserves the time reversal symmetry, such as $Cu_xBi_2Se_3$ [13]. MF are supposed to be contained by states are surface of topological superconductors. From Dirac Fermions they are quite different, and they are the particle having separate antiparticle, though MF itself are antiparticles. In materials discovery of TSC answered the realization of MF [14].

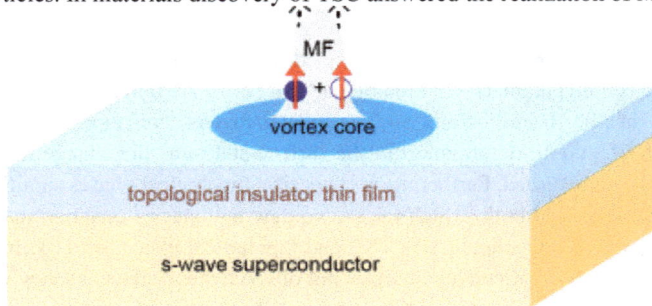

Figure 1. Visual representation of TSCs [15]

To obtain topological superconductivity several methods are available such as applying pressure to TIs, doping and applying the proximity effect [16, 17]. Topological superconductors are positioned in proximity of s-wave superconductor, in proximity based TSC, and superconductivity is shown by Dirac fermions positioned on the TI's surface. In TIs doping of certain elements is another technique to persuade superconductivity. First TSC based on doping came into existence when superconductivity was shown by Bi_2Se_3 because of Cu intercalation among its two quintuple layers. Actually, there are not sufficient carriers in present in Bi_2Se_3 to show superconductivity, so intercalation of Cu enhances the amount of charge carriers. This discovery leads to the advance research for doped TSC. Furthermore it was found that superconductivity was shown through pressure in some TIs, for example superconductivity was observed in Bi_2Te_3 close to 2.7K at 7GPa pressure [18].

2.   Theory of topological superconductors

In super conducting states, BdG Hamiltonian's all negative states are fully occupied. So, to define topological numbers for each state is easy similar to topological insulators. Different topological numbers can be introduced that depends upon symmetry and dimensions of the system. Broadly, to make the superconductor topological any of such topological number should be non-zero. Nodal superconducting gap is often found in unconventional superconductors, where each node has topological number. Hence such unconventional SCs are topological and might be taken as weak TSCs. In a narrower sense, for topological superconductivity complete open gap is needed other than non-zero topological number. In such case, the system is termed as strong TSCs as no gap-less bulk excitations are found. Similar to QH states, at low temperature, the transport properties of such system is determined through gapless excitations protected topologically restricted at topological defect or edge [19].

3.   Majorana fermions

To realize Majornana fermions the best platform is TSCs [9]. Though they are not detected yet, but still TSCs fascinate both experimental and theoretical research. In 1929, Dirac equations were derived by Paul Dirac for relativistic motion. It is predicted by theory that there exist an antiparticle for each fermion that is different in charge and sign. In 1934, this prediction was confirmed experimentally, when new particle Positron was searched by Anderson whose mass and spin was same as electron but sign was opposite. In 1973, presence of such particles was predicted by ATTORE Majorana which was identical to their antiparticle in all characteristics; to made this prediction Dirac equation was used. He drives the wavefunction's complex conjugate in his work, which satisfies the Dirac equation, without any incongruity will also satisfy the same. It is considered that the wavefunction's complex conjugate that satisfies the Dirac equation is the antiparticles wavefunction. The existence of particles same as their antiparticle if wavefunction is real and would be self-conjugate. These particular kinds of particles that are antiparticles of their own are Majorana particles [20]. Fermi Dirac statistics followed by these particles, that's the reason which makes them Majorana fermions as well, the debate took place after theoretical prediction that which system will be used to realize Majorana fermions. According to particle physics, conditions to be the Majorana fermions is only full filed by neutrinos, but still it is argumentized that whether they can be taken as Majorana fermions. Two conditions need to be satisfied to be the Majorana fermions one is that it should be its own antiparticle and the other is to satisfy the Dirac equation. Electron does not satisfy these conditions as in condensed matter it is integral fermion and the non-relativistic Schrodinger equation obeys by them along with negative charge, so this eliminates very possibilities for it to be its own antiparticle [21]. Thus, superconducting topological materials satisfies all conditions to be Majorana fermions apart from neutrinos. Two factors give the possibility to realize the Majorana fermions: 1) To host excitations TSCs have topological non-trivial surface states, directed through Dirac equation. 2) Presence of superconductivity. Electron-hole excitation superimposed in TSCs which makes them charge-neutral and indistinguishable which allows them to be their own antiparticles. Hence, for the experimental realization of Majorana fermions TSCs are very beneficial materials [22].

4.   Possible candidate of superconductivity in TSCs

Superconductivity in TSCs is a most developing area of study. Topological superconductivity initially proposed at the s-wave superconductor and at interface of topological super conductors theoretically. Later on, in some non-centrosymmetric and unconventional superconductors were fund to have topological superconductivity in its intrinsic form. Carrier doping of specific dopants can also induce superconductivity in few topological materials, and with applied pressure this phenomenon can also be observed. In heterostructures contains SCs and TIs or SCs and high spin orbit coupling materials TSCs can be observed [22].

4.1   Unconventional superconductors (SCs)

Full superconducting gap having odd symmetry pairing must be shown by one compound to be the potential candidate of TSCs, but s-wave SCs (conventional SCs) are the known SCs [1]. Spin singlet pairing is found in SCs and are topological trivial mostly. Hence to observe superconductivity topological in nature in the pure form, topological non-trivial disposition along with unconventional pairing proportion must be found in SCs. Spin-triplet SCs could give the glimpse of unconventional SCs for example cooper pairs, they have S=1 spin angular momentum. Such SCs can only be paired with L=1,3,5... orbital angular momentum and p,d... orbitals. The dissimilarity among spin triplet and singlet pairing is that paramagnetic susceptibility in spin triplet paring is maintained while in spin singlet pairing paramagnetic susceptibility is reduced in cooper pairs. Classification of unconventional SCs can be done as p and d-wave SCs in accordance with pairing symmetry. The prediction is that Majorana fermions host the vortex core of p-wave SCs [21]. These special kind of SCs are supposed to be possible candidates for TSCs. Magnetic field can break the pairing easily, this is main issue with this class of SCs. Sr_2PuO_4 is an most studied example of such class of SCs [22].

4.2   Iron based superconductors

Topological non-trivial surface states are found in iron based SCs for example LiFeAs and $FeSe_{0.5}Te_{0.5}$ in ARPES measurements and first principal calculations. But there still has not been any proof of bulk topological properties. Kamihara et.al. discovered iron based SCs in $La[O_{1-x} F_x]FeAs$ in 2008 [23], since then it has been hot topic for research. Superconductivity has been shown by FeSe monolayer grown on substrate of $SrTiO_3$ near 100K, over seven year since they have discovered, and to study high Tc superconductivity made this system (FeSe/STO) an exemplary system [24]. On $FeTe_{0.55}Se_{0.45}$ STS measurements was performed, in vortex core sharp zero biased conduction peak was observed, it remained at zero energy and did not split when moved away from vortex center. Majorana bound states was observed in vortex core. On the monolayer of $FeTe_{0.55}Se_{0.45}$ STS measurements was performed in which the zero energy bound states appear and made it possible candidate for TSC. In the recent study observation of zero energy modes takes place on unit cell of $FeTe_{0.55}Se_{0.45}$ and FeSe deposited on substrate of STO [24]. In the topological superconductivity frame work iron arsenide system established as ineluctable system through these results. To consider this system TSC, still more experimentation is required on this system. It is predicted that some arsenide based (CaAs and $CaFeAs_2$) and heavy electron doped FeSe ($Li_{0.84}Fe_{0.16}OHFeSe$) materials could be candidates of TSCs [25, 26].

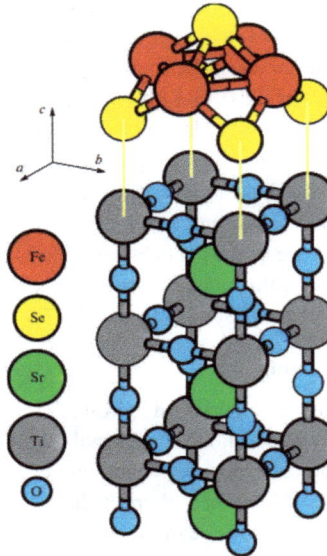

Figure 2. FeSe monolayer crystal structure[27]

4.3   Tin based superconductors

Superconductivity was found in SnTe having specific dopant, while there are different Sn-based systems, for TSCs they are possible candidate for example SnSb alloy. Long ago at 2.3K superconductivity was found in SnSb. It has n-type carriers and is a type II superconductor. But still significantly its superconductivity has not explored. SnSb in specific heat measurements it was found as complete gapes s-wave superconductor. Antimony is known for having topological properties [28].

SnAs is another tin-based superconductor, for topological superconductivity it is regarded as good candidate. SnAs has 3.5K-4K range of critical temperature and its cubic structure is similar to NaCl. Superconductivity with s-wave pairing is conventional in SnAs. But superconductivity' nature is uncertain in SnAs. To explore SnAs for superconducting properties along with topological properties it is an interesting system [28].

Sn_4As_3 for topological superconductivity it is a potential candidate. It has 1.1K Tc and is non-centrosymmetric SC. ARPES measurements providing the evidence of surface state in

Sn$_4$As$_3$ [29]. To explore its topological properties more work needs to be done on this compound.

5. Properties of topological superconductors

5.1 Spin current and thermal conductivity

Through transport experiments gapless surface modes could be searched. However, due to superconducting states's zero resistivity DC charge transport measurements are not convenient, so to study superconducting state, spin transport or thermal transport can be employed [30]. It is noted that heat cannot be carried by cooper pairs and by T^3 law phonon die away, which makes the Majorana modes only heat carriers at low temperatures. Hence, indication of Majorana fermions, in a complete gapped superconductor is given by finite surface thermal conductivity. Moreover, a heat current corresponds with spin current when helical spin polarization is given by surface modes for example Cu$_x$Bi$_2$Se$_3$. Measurements of spin polarization induced by heat current on the surface of TSC provides further confirmation of Majorana fermions [19].

5.2 Anomalous Josephson effect

Josephson junctions among s-wave superconductor and time-reversal invariant TSCs in the junction comprise surface Majorana modes, which results to an anomalous current-phase correlation. This is essential due to phase difference of $\theta = 0, \pi$ Majorana modes are gapless but gap appears for $\theta \neq 0, \pi$ that takes it to the half-period Fraunhofer pattern [31]. When 3D time reversal invariant TSC is inserted into the c-shaped s-wave SC to produce continuous rings, it is predicted that in the ring flux quantization in half units of the flux quantum h/4e, for the similar reason. Interestingly if d-wave SC replaces the C-shaped portion of ring, h/2e(n+1/2) value taken by flux quantization with integer n. To the symmetry of the system the Josephson current phase equation is sensitive. The decomposition of Josephson into a harmonic series as such:

$$J(\theta) = \sum_{n=1}(J_n \sin n\theta + I_n \cos n\theta) \qquad (1)$$

Where J_n and I_n vanishes as n increases. $J(\theta)$ transforms to $-J(-\theta)$ under time reversal operation which means for the event of time reversal changeless junction, $I_n = 0$. If the mirror reflection arrangement is preserved by junction, then current phase association also depend on gap function' mirror parity. When junction is considered among a conventional s-wave SC and mirror odd SC, mirror reflection symmetry suggests $J(\theta) = J(\theta + \pi)$ which gives J$_{2n+1}$=0. As a result, Josephson current obtained having anomalous current-phase relationship $J(\theta) \sim \sin 2\theta$. For the identification of TSC's bulk pairing symmetry such an anomalous relation is helpful [19].

5.3   Majorana fermions in hybrid systems

In hybrid system to engineer artificial TSCs to prove the existence of Majorana fermions various predictions regarding their unusual properties have been made. Some of bulging predictions are 4π periodic Josephson effect, braiding Phenomena related to Majorana exchange, at a ballistic NS point contact junction half integer conductance quantization, interference effect around a sea of Majorana fermions, Majorana teleportation [19, 32].

5.4   Nematicity

The Δ_4 states realized by $Cu_xBi_2Se_3$ which can be assumed as nematic SC [33]. In a Knight shift, upper critical field, on the orientation their dependency in magnetic field and specific heat, detection of nematicity has been done by uniaxial anisotropy. In similar SCs the existence of nematicity has been confirms such as $Nb_xBi_2Se_3$ and $Sr_xBi_2Se_3$ [34]. In many different properties nematicity should evident itself such as elastic constant, thermal conductivity, sound velocity, penetration depth [19].

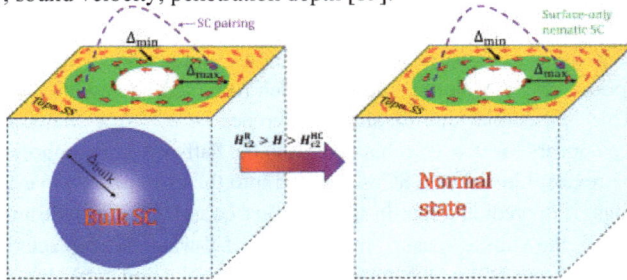

Figure 3. Graphical representation of nematicity [35]

Conclusion

TSCs have gained much of interest recently and new materials have been explored that can show topological superconductivity, though different materials haven been explored theoretically but experimental verification still is in process. Topological superconductivity has direct relation with Majorana fermions and sign of their presence has been established. TSCs are important as they host Majorana fermions in the bulk form. This chapter covers the detailed discussion of Majorana fermions and concept of TSCs. In doped TIs along with bulk SCs, topological superconductivity is convenient way for the realization of topological superconducting states. Pressure induction is another method to realize topological superconductivity in TSCs. Suited materials and properties of TSCs has been discussed briefly.

References

[1] X.-L. Qi, S. Zhang, Topological insulators and superconductors, Reviews of Modern Physics 83 (2011) 1057. https://doi.org/10.1103/RevModPhys.83.1057

[2] M.Z. Hasan, C.L. Kane, Colloquium: Topological insulators, Reviews of Modern Physics 82 (2010) 3045. https://doi.org/10.1103/RevModPhys.82.3045

[3] K.V. Klitzing, G. Dorda, M. Pepper, New method for high-accuracy determination of the fine-structure constant based on quantized hall resistance, Physical Review Letters 45 (1980) 494-497. https://doi.org/10.1103/PhysRevLett.45.494

[4] S. Zhang, Topological states of quantum matter, Physics 1 (2008) 6. https://doi.org/10.1103/Physics.1.6

[5] C.L. Kane, E.J. Mele, Quantum spin Hall effect in graphene, Physical Review Letters 95 (2005) 226801. https://doi.org/10.1103/PhysRevLett.95.226801

[6] M. König, S. Wiedmann, C. Brüne, A. Roth, H. Buhmann, L.W. Molenkamp, X.L. Qi, S.C. Zhang, Quantum spin hall insulator state in HgTe quantum wells, Science 318 (2007) 766-770. https://doi.org/10.1126/science.1148047

[7] B.A. Bernevig, S.C. Zhang, Quantum spin Hall effect, Physical Review Letters 96 (2006) 106802. https://doi.org/10.1103/PhysRevLett.96.106802

[8] A. Roth, C. Brüne, H. Buhmann, L.W. Molenkamp, J. Maciejko, X.L. Qi, S.C. Zhang, Nonlocal transport in the quantum spin Hall state, Science 325 (2009) 294-297. https://doi.org/10.1126/science.1174736

[9] C. Nayak, S.H. Simon, A. Stern, M. Freedman, S. Das Sarma, Non-Abelian anyons and topological quantum computation, Reviews of Modern Physics 80 (2008) 1083-1159. https://doi.org/10.1103/RevModPhys.80.1083

[10] M. Sato, Y.J.R.o.P.i.P. Ando, Topological superconductors: a review, 80 (2017) 076501. https://doi.org/10.1088/1361-6633/aa6ac7

[11] Y. Ando, L.J. Fu, Topological crystalline insulators and topological superconductors: From concepts to materials, Annual Review of Condensed Matter Physics 6 (2015) 361-381. https://doi.org/10.1146/annurev-conmatphys-031214-014501

[12] K. Ishida, H. Mukuda, Y. Kitaoka, K. Asayama, Z. Mao, Y. Mori, Y.J.N. Maeno, Spin-triplet superconductivity in Sr2RuO4 identified by 17O Knight shift, Nature 396 (1998) 658-660. https://doi.org/10.1038/25315

[13] L. Fu, E.J. Berg, Odd-parity topological superconductors: Theory and application to CuxBi 2Se3, Nature 105 (2010) 097001. https://doi.org/10.1103/PhysRevLett.105.097001

[14] P. Sharma, N. Karn, V.S. Awana, Technology, A comprehensive review on topological superconducting materials and interfaces, arXiv:2204.12107 (2022). https://doi.org/10.1088/1361-6668/ac6987

[15] H.H. Sun, J.F. Jia, Majorana zero mode in the vortex of an artificial topological superconductor, Science China Physics, Mechanics & Astronomy 60 (2017) 057401. https://doi.org/10.1007/s11433-017-9011-7

[16] L. Fu, C.L. Kane, Superconducting proximity effect and Majorana fermions at the surface of a topological insulator, Physical Review Letters 100 (2008) 096407. https://doi.org/10.1103/PhysRevLett.100.096407

[17] J.D. Sau, R.M. Lutchyn, S. Tewari, S.D. Sarma, Generic new platform for topological quantum computation using semiconductor heterostructures, Physical Review Letters 104 (2010) 040502. https://doi.org/10.1103/PhysRevLett.104.040502

[18] K. Matsubayashi, T. Terai, J. Zhou, Y.J. Uwatoko, Superconductivity in the topological insulator Bi2Te3 under hydrostatic pressure, Physical Review B 90 (2014) 125126. https://doi.org/10.1103/PhysRevB.90.125126

[19] M. Sato, Y. Ando, Topological superconductors: A review, Reports on Progress in Physics 80 (2017) 076501. https://doi.org/10.1088/1361-6633/aa6ac7

[20] T. Angsachon, R. Dhanawittayapol, K. Kritsarunont, S.N. Manida, A free solution to the Dirac equation in R-spacetime, Journal of Physics: Conference Series 1380 (2019) 012090. https://doi.org/10.1088/1742-6596/1380/1/012090

[21] F.J.N.P. Wilczek, Majorana returns, Nature Physics 5 (2009) 614-618. https://doi.org/10.1038/nphys1380

[22] M.M. Sharma, P. Sharma, N.K. Karn, V.P.S. Awana, Comprehensive review on topological superconducting materials and interfaces, Superconductor Science and Technology 35 (2022) 083003. https://doi.org/10.1088/1361-6668/ac6987

[23] Z. Wang, P. Zhang, G. Xu, L.K. Zeng, H. Miao, X. Xu, T. Qian, H. Weng, P. Richard, A.V. Fedorov, H. Ding, X. Dai, Z. Fang, Topological nature of the

FeSe0.5Te0.5 superconductor, Physical Review B 92 (2015) 115119.
https://doi.org/10.1103/PhysRevB.92.115119

[24] Y. Kamihara, T. Watanabe, M. Hirano, H. Hosono, Iron-based layered superconductor La[O1-xFx]FeAs (x = 0.05−0.12) with Tc = 26 K, Journal of the American Chemical Society 130 (2008) 3296. https://doi.org/10.1021/ja800073m

[25] D. Li, Y. Liu, Z. Lu, P. Li, Y. Zhang, S. Ma, J. Liu, J. Lu, H. Zhang, G. Liu, F. Zhou, X. Dong, Z. Zhao, Quasi-two-dimensional nature of high-Tc superconductivity in iron-based (Li, Fe)OHFeSe, Chinese Physics Letters 39 (2022) 127402. https://doi.org/10.1088/0256-307X/39/12/127402

[26] Z.T. Liu, X. Xing, M.Y. Li, W. Zhou, Y. Sun, C. Fan, H.F. Yang, J.S. Liu, Q. Yao, W. Li, Z.X. Shi, D. Shen, Z.J.a.S. Wang, Observation of the anisotropic Dirac cone in the band dispersion of 112-structured iron-based superconductor Ca0.9La0.1FeAs2, Applied Physics Letters 109 (2016) 042602. https://doi.org/10.1063/1.4960164

[27] I.A. Nekrasov, N.S. Pavlov, M.V. Sadovskii, Electronic structure of FeSe monolayer superconductors: Shallow bands and correlations, Journal of Experimental and Theoretical Physics 126 (2018) 485. https://doi.org/10.1134/S1063776118040106

[28] S. Geller, G.J. Hull Jr, Superconductivity of intermetallic compounds with NaCl-type and related structures, Physical Reviwe Letters 13 (1964) 127. https://doi.org/10.1103/PhysRevLett.13.127

[29] C.A. Marques, M.J. Neat, C.M. Yim, M.D. Watson, L.C. Rhodes, C. Heil, K.S. Pervakov, V.A. Vlasenko, V.M. Pudalov, A.V. Muratov, T.K. Kim, P. Wahl, Electronic structure and superconductivity of the non-centrosymmetric Sn4As3, New Journal of Physics 22 (2020) 063049. https://doi.org/10.1088/1367-2630/ab890a

[30] A.P. Schnyder, S. Ryu, A. Furusaki, A.W.W. Ludwig, Classification of topological insulators and superconductors in three spatial dimensions, Physical Review B 78 (2008) 195125. https://doi.org/10.1103/PhysRevB.78.195125

[31] S.B. Chung, J. Horowitz, X.L. Qi, Time-reversal anomaly and Josephson effect in time-reversal-invariant topological superconductors, Physical Review B 88 (2013) 214514. https://doi.org/10.1103/PhysRevB.88.214514

[32] J. Alicea, New directions in the pursuit of Majorana fermions in solid state systems, Reports on Progress in Physics 75 (2012) 076501. https://doi.org/10.1088/0034-4885/75/7/076501

[33] K. Matano, M. Kriener, K. Segawa, Y. Ando, G.Q. Zheng, Spin-rotation symmetry breaking in the superconducting state of CuxBi2Se3, Nature Physics 12 (2016) 852. https://doi.org/10.1038/nphys3781

[34] S. Yonezawa, K. Tajiri, S. Nakata, Y. Nagai, Z. Wang, K. Segawa, Y. Ando, Y.J.N.P. Maeno, Thermodynamic evidence for nematic superconductivity in CuxBi2Se3, Nature Physics 13 (2017) 123. https://doi.org/10.1038/nphys3907

[35] T. Le, Y. Sun, H.K. Jin, L. Che, L. Yin, J. Li, G. Pang, C. Xu, L. Zhao, S. Kittaka, T. Sakakibara, K. Machida, R. Sankar, H. Yuan, G. Chen, X. Xu, S. Li, Y. Zhou, X. Lu, Evidence for nematic superconductivity of topological surface states in PbTaSe2, Science Bulletin 65 (2020) 1349. https://doi.org/10.1016/j.scib.2020.04.039

Topological Insulators: Materials and Applications   Materials Research Forum LLC
Materials Research Foundations 154 (2024) 95-119   https://doi.org/10.21741/9781644902851-6

Chapter 6

Manganese Doped Topological Insulators

M.W. Yasin and S.S. Ali*

School of Physical Sciences, University of the Punjab Lahore, 54590, Pakistan

shahbaz.sps@pu.edu.pk

Abstract

When a topological insulator is incorporated with magnetism, the time-reversal symmetry (TRS) breaks. This chapter sums up the current research in Mn-doped topological insulators, mostly focusing on antiferromagnetic $MnBi_2Te_4$ and its family. Specific critical behavior of Mn-doped Bi_2Te_3 topological insulator reveals ferromagnetic ordering (to increase conductivity) of Mn spin at normal substitution. Moreover, due to intrinsic anti-state substitutional defect if a comparable amount of local moment is produced then calculated data is in favor of a spin glass state with polarization. Exotic phenomena like the magnetoelectric effect and dissipation less edge state occur in this material and to study $MnBi_2Se_4$ epitaxial thin films a scanning tunneling microscope (STM) is utilized. Steps between a screw dislocation and the van der Waals layer are manifested in the large-scale topographic image. In a small area at the edge Bi termination is occurred and Se termination is dominant at the surface, this termination is compared by examining step height, tunneling spectroscopy, and resolution image. Similarly, for analyzing Mn impurity in Bi_2Se_3 electron paramagnetic resonance is used and by implementing the vertical Bridgman method topological insulator is grown. Mn^{2+} configuration was found in conducting state of a host metal and Mn in high spin $= 5/2Mn^{2+}$. This assumption shows that energy level $Mn^{2+}(d^5)$ is present the valence bad (VB) and energy level, $Mn^{1+}(d^6)$ is located far way form energy gap.

Keywords

Dirac Point, Ferromagnetic, Quintuple Layer, Quantum Anomalous Hall Effect, Susceptibility and Photoemission Intensity

Abbreviation

TIs (topological insulators), AFM (antiferromagnetism), STM (scanning tunneling microscope), VB (valence band), CB (conduction band), RKKY (Rudermen-Kittle-

Topological Insulators: Materials and Applications
Materials Research Foundations 154 (2024) 95-119

Materials Research Forum LLC
https://doi.org/10.21741/9781644902851-6

Kazuya-Yosdia), MBT (*MnBi₂Te₄*) and QAHE (quantum anomalous hall effect), time-reversal symmetry (TRS)

Contents

1.   Introduction

At the Dirac point topological band gap opening and surface state are modified by impurities, this novel phenomenon occurs when the topological insulator is doped with a transition metal. High spin electronic configuration and highest solubility of all transitions make it the most popular dopant and RKKY (Rudermen-Kittle-Kazuya-Yosdia) interaction with localized magnetic moment mediated by Dirac state and long-range ferromagnetic order occur as a result. Recently, in Mn-doped molecular beam epitaxy-grown BiSe$_3$, ferromagnetic behavior was observed [1,2]. The method by which Mn is incorporated in the topological insulator crystal has not been fully explored yet and for the clarification of the origin of magnetic interaction, it is compulsory to analyze the spin state of impurity and charge. In Bi$_2$Se$_3$, the spin state of Mn is based on the Fermi level but still experimentally not investigated, normally impurity serves as an acceptor band in Mn. BiSe$_3$ has a 2+ valence state, on the basis of the Mn-doping effect in the electric properties of Bi$_2$Se$_3$ and the acceptor character has been analyzed [3].

Employing an external magnetic field time reversal regularity of conducting surface symmetry overwhelmed resulting in an energy gap at the Dirac point. Exotic effects like the Hall effect and charge carrier tunning can be produced by this process. For a better understanding of fabrication and electronic properties, we need to know an electronic site. In mostly 3D TIs-like tetradynmites (Bi$_2$S$_3$ and Bi$_2$Te$_3$) maintaining a stoichiometric ratio is the most difficult task. Although for the construction of a potential device elegant tunning of charge carrier is required [4]. A category of quantum materials that are distinguished by gapless surface states is represented by TIs. Doping of TI with magnetic ions is the finest way to investigate the effect of breaking TRS and this breaking symmetry induced many topological states [5].

Topological insulators state classified by a new type that is based on topological order and is explained by a global quantity which is not suitable for breaking the symmetry paradigm. More accurately, an unusual topological property is possessed by the group of VBs (valence bands). A band insulator is an insulator whose set of conduction and valence band is separated by an energy gap (E_g) [6]. All insulting phases are equivalent to each other is a fundamental question rise in the topological classification of insulators. The topological index is a perfect example of a two-dimensional manifold Euler-Poincare characteristic. The presence of a gapless surface state or edge is an extraordinary result of bulk nontrivial topology, the topological insulator surface must be metallic [7].

2. Structure

Bi$_2$Se$_3$ unit cells contain three quintuple layers and crystallize into a rhombohedral lattice. Five interchanging Bi and Se hexagonal planes are present in each layer gathering with the c-axis according to the atomic order (Se − Bi − Se − Bi − Se). Selenium atoms present in the outermost layer of the quintuple and quintuple layers are bonded with Van der Waals forces having Van der Waal gap produced between them. Multiple selenium vacancy defects behave as double donor V_{se}. Therefore, pure Bi$_2$Se$_3$ is generally n-type, and the Bridgeman growth method is used to control electron concentration by adjusting the stoichiometry of the melt because pure Bi_2Se_3 electron concentration greatly depends upon the growth conditions [8]. The highest electron concentration (10^{19}-10^{20} cm^{-3}) is obtained in crystal grown at room temperature and concentration can be reduced to 10^{17} cm^{-3} by increasing the ratio of selenium to bismuth. When Mn is doped in Bi_2Se_3 conduction converts into P-type because doped Bi_2Se_3 acts as an acceptor [3,9].

By using stoichiometric melt and a high bismuth-selenium-to-magnese atomic ratio (Bi: Se: Mn = 1.85: 3.00: 0.15), the first experiment was done to dope Mn in Bi_2Se_3 at $25°C$ (room temperature). The n-type ingot (material) was obtained with an electron concentration of 1.1×10^{20} cm^{-3}. Initially, it was suggested that Mn reduced electron concentration and this antisite defect behaves as an acceptor, but recently it was declared that Bi_{se} defect is a donor present in the outermost Se layer observed as V_{se} and Be interstitial [10]. Next crystallization experiment was performed by moderate Se excess and excellent crystal morphology is obtained. Free electron concentration of pure material for particular stoichiometry and growth condition of Mn-doped Bi_2Se_3 is represented in Table 1 [11]. A slight excess of Se is present (Bi: Se: Mn = 1.80: 3.14: 0.050) in ingots acquired from melt and p and n-type regions developed with ingot. The concentration of electrons in n and p-type region is 9×10^{18} cm^{-3}, (6-9)$\times 10^{18}$ cm$^{-3,}$ and 1.5×10^{-9} cm^{-3} concentration of Mn^2 present in both types. Mn amount increased in melt from 3.6×10^{18} cm^{-3} to 1.1×10^{19} cm^{-3} for atomic part 0.015 to 0.100 of Mn, hole concentration consistently increases, and p-type conductivity attained by balanced Se amount in melt (Bi : Se = 1.6: 3.30) [3].

Table 1. Mn-doped Bi_2Se_3 sample produced by the Bridgeman method, free electron concentration in pure crystal and concentration of Mn^{2+} calculated from EPR. Reused from [3].

Ingot number	Nominal melt Composition(atolmic ratio)		Comments	Carrier concentration(cm^{-3})		Refrence	Mn^{2+}	Morphology
	Bi:Se	Mn		n	p	n(cm^{-3})		
1	1.85:3.00	0.150 7.5 at%	Stoichiometric melt	1.1×10^{20}		2×10^{18}	No EPR signal	Tail:BiSe, Se
2	1.80:3.15	0.050 2.5 at%	Slight Se excess n/p regions	8.8×10^{18}	5.8×10^{18}	5×10^{18}	1.5×10^{18}(cm^{-3}) 0.11 at%	No data
3				8.6×10^{18}	8.8×10^{18}		No EPR signal	No data
4	1.60:3.30	0.100 5 at%	Moderate Se excess	1.1×10^{18}	1×10^{18}			
5	1.65:3.30	0.049 2.45 at%			6.4×10^{18}		1.2×10^{19}cm^{-3} 0.087 at%	
6	1.68:3.30	0.015 0.75 at%			3.6×10^{18}		3.9×10^{18}cm^{-3} 0.028 at%	
7	1.70:3.30	0.008 0.4 at%	Moderate Se excess +Ca doping(0.008)	4.6×10^{18}		No data	2.4×10^{18}cm^{-3} 0.017 at%	
8	1.69:3.30	0..0037 0.185 at%	Moderate Se excess +Ca doping(0.0035)	6.0×10^{18}		No data	No EPR signal	

The nanocrystal of $Bi_{2-x}Mn_xTe_3$ (topological insulator) heated in the glass tube at 500°C for 10 hours is manifested by transmission electron microscope as represented in Fig. 1. The interplanar distance and average size of the nanocrystal are 0.322 nm and 0.5 nm with crystalline hexagonal phase (015) of Bi_2Te_3 as analyzed by imaging software [12-14].

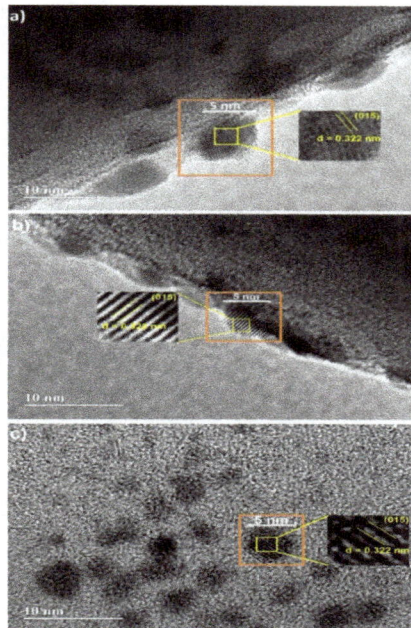

Figure 1. Transmission electron microscope image of Mn doped Bi_2Te_3 at different concentrations $X_{Mn}=0.00$, 0.05, and 0.10 with a size around 0.5 nm and the crystalline plane is (015). Reused from [12].

2.1 Layered structure of MnBi₂Se₄

Mn-doped Bi_2Se_4 is a layered material composed of septuple layer (SL) distinct by Van der Waals gap, having space group R3m as represented in Fig. 2, grouping order is $Se - Bi - Se - Mn - Se - Bi - Se$. The lattice constant is between 3.93 Å and 4.22 Å calculated by DFT (density functional theory) and each septuple layer consists of triangular lattice atoms, the distance between SL layer is 12.26-13.05 Å [15-16]. To study the morphology of complete film Fig. 2 displays extensive topographic STM images in which peninsulas are seen in the topographic image that led to height deviation over the mage almost ±1 septuple layer and smooth atomically flat areas are also observed. The highlighted red circle shows the sloping across the surface in the atomically flat areas due to the presence of screw dislocation (SD). On further investigation, it is found that generally all surfaces are smoothly attached because of screw dislocations, which suggested that these areas must have similar surface termination. The height location indicated with the three lines in Fig. 2(c) is an intensified or magnified version of a square box in Fig. 2(b). The red profile along the top of the peninsula, the blue profile with the right edge of the peninsula and the green profile near screw dislocation have approximate heights of 1.6 Å, 12.2 Å and 6 Å [17].

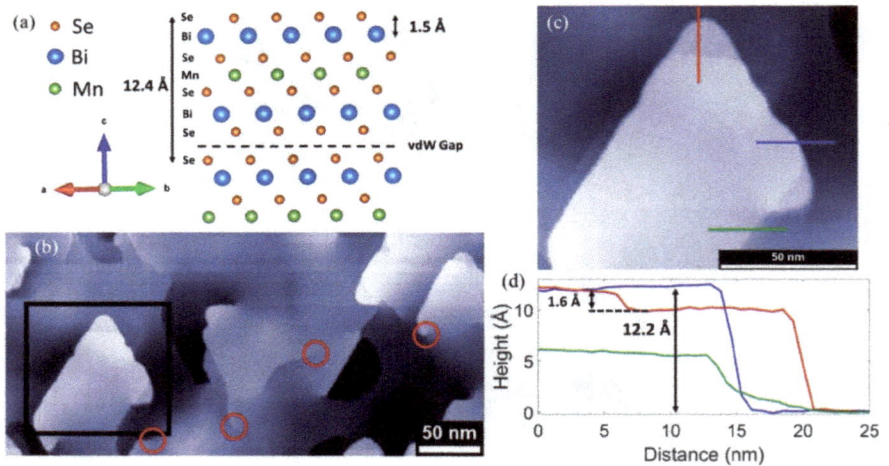

Figure 2. (a) Portrait view of $MnBi_2Se_4$ single crystal,(b) topographic scanning tunneling image at a voltage -2V and current 0.2 nA, some screw dislocations are manifested by red circles. The box area in (b) is zoomed in (c). Height profile Red, blue, and green line are explained in (d). Reused from [17].

2.2   Vapor transport growth of $MnBi_2Te_4$

To grow $MnBi_2Se_4$, halides were used as transport agents. MnI_2, $TeCl_4$, or $MoCl_5$ were utilized as transport agents for the growth of $MnBi_2Te_4$ and as starting material premelted $Mn_2Bi_2Te_5$ used by keeping in mind the slow transportation of Mn [18]. When I_2 and MnI_2 were used as transport agents, no difference was observed in the growth of $MnBi_2Te_4$

crystal, although these are more effective transport agents compared to chlorides. Table 2 shows that a longer time is required to get crystals of comparable size when a chloride transport agent is used. A small amount of temperature < 20°C is required for perfect growth of $MnBi_2Te_4$ and related growth as illustrated in Fig. 2(a). When the temperature gradient is greater than 20°C, 1-4% Mn is doped in Bi_2Te_4 then a small amount of MnTe crystal is found at the growth container hot end. Crystals-like plate shapes are obtained with any type of transport agent when the experiment is performed inside a furnace tube (see Fig. 3(c)). In contradiction, when a box furnace is used crystals with rectangular bar shapes and plate shapes are produced inside the growth chamber (see Fig. 3(b)). After the growth of two different samples in a box furnace using transport agent I_2 picture of the cold end is shown in Fig. (d,e). Mostly MBT ($MnBi_2Te_4$) crystals remain in the cold end, sometimes found between hot and cold ends across the tubes. All these situations suggest that temperature in a tube furnace is well-defined compared to a box furnace, although all growth inside a box furnace like $MnBi_{2-x}Te_4$, $MnSb_2Te_4$, and $MnBi_2Te_4$ are reproducible [19,20].

Table 2. Some selected growths, for last four batches only Sb content, x in $Mn_2Bi_{2-x}Sb_xTe_5$ is written down in starting composition. Single tube furnace used for growth with additional thermocouple to control cold end temperature. A cusp was noticed at Neel temperature T_N, from hall measurement at room temperature, concentration and charge carrier type was determined. Reused from [19].

Batch #	Starting materials	Agent	T/time	Furnace	EDS	$T_{N/C}$	e/h ($\times 10^{19}$ cm^{-3})
#1164	Mn:Bi:Te = 2:2:5	I_2	585 °C, 10 days	box	$Mn_{0.94(2)}Bi_{2.07(1)}Te_{3.99(1)}$	25.0 K	e, 3.1
#1174B	Mn:Bi:Te = 2:2:5	$TeCl_4$	585 °C, 14 days	box	$Mn_{0.94(2)}Bi_{2.05(2)}Te_{4.01(3)}$	24.6 K	e.7.3
#1178	Mn:Bi:Te = 2:2:5	$MnCl_2$	585 °C, 21 days	box	$Mn_{1.01(2)}Bi_{1.99(2)}Te_{4.00(2)}$	25.2 K	e. 5.0
#1183B	Mn:Bi:Te = 5:2:8	I_2	550 °C, 14 days	tube	$Mn_{1.01(1)}Bi_{2.01(1)}Te_{3.98(1)}$	25.7 K	e. 3.3
#1198B	Mn:Bi:Te = 2:2:5	I_2	550 °C, 14 days	tube	$Mn_{0.99(1)}Bi_{2.00(1)}Te_{4.01(1)}$	25.7 K	e, 1.2
#1193A	Mn:Sb:Te = 2:2:5	I_2	590 °C, 14 days	box	$Mn_{1.15(2)}Sb_{1.95(1)}Te_{3.90(1)}$	$T_C = 44$ K	h.110
#1191A	x = 0.65	I_2	590 °C, 14 days	box	$Mn_{1.00(4)}Bi_{1.34(3)}Sb_{0.74(3)}Te_{3.92(3)}$	24.2 K	–
#1238	x = 1.0	I_2	585 °C, 14 days	box	$Mn_{0.97(4)}Bi_{1.04(1)}Sb_{1.05(2)}Te_{3.94(1)}$	23.5 K	–
#1193B	x = 1.5	I_2	590 °C, 14 days	box	$Mn_{1.01(3)}Bi_{0.50(3)}Sb_{1.53(1)}Te_{3.95(1)}$	$T_C = 26$ K	–
#1193C	x = 1.0	I_2	590 °C, 12 days	box	$Mn_{0.98(1)}Bi_{1.77(1)}Sb_{2.39(1)}Te_{4.96(1)}$	$T_C/T_N = 13,$ 6 K	h.37

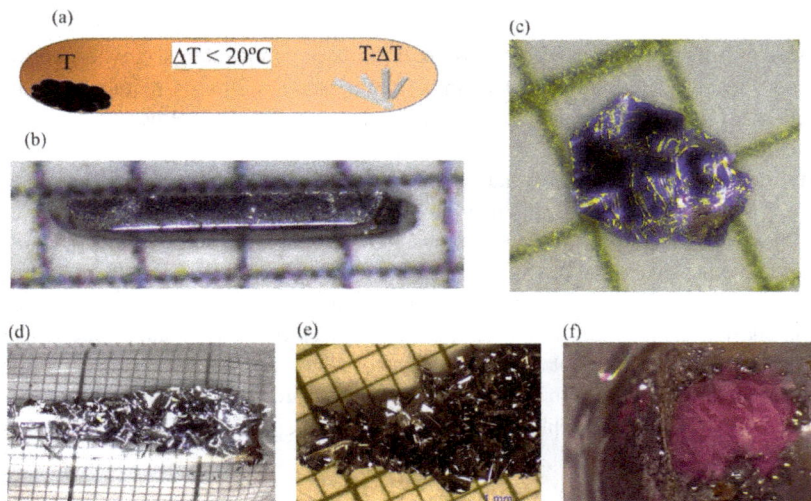

Figure 3. (a) Chemical vapor transport growth schematic picture at a gradient temperature less than 20ºC for outstanding growth of MnBi₂Te₄, whereas Fig. 3(b) illustrates a rectangular-bar-shaped crystal of MnBi₂Te₃. MnBi₂Te₃ plate shape crystal is presented in Fig. (c). In box furnace, at the cold end, crystal like rectangular bar-shaped and plate shape form as illustrate in Fig. 3(d). When transport agent I₂ used cluster crystal grown manifested in Fig. 3(d) and Fig. 3(f) show beautiful MnCl₂ crystal and droplet of Bi₂Te₃ melt which is condensed at cold end. Reused from [17].

In inclusion to a crystal of favored composition, inside of chamber following crystals are generally found, Mn-doped Bi_2Te_3, MnTe, $MnTe_2$, and MnX where X=Cl, I, here $MnCl_2$ crystals are moisture sensitive and transparent. To get a perfect crystal for observing quantum anomalous Hall effect transport vapour growth has been studied using $MnCl_2$ as a transport agent. $MnBi_2Te_4$ crystals are utilized in the device to display the quantum anomalous Hall effect (QAHE). $MnBi_2Te_4$ is only responsible for QAHE and the growth of the crystal in the existence of $MnCl_2$ required full attention. The color of I_2 appears when transport agent MnI_2 is used but the color suddenly disappears, when an ampoule growth is inserted in the furnace, this purposed that the reaction of I_2 and decomposition of MnI_2 formed a new phase with high vapor pressure. Firstly, vapor growth of $MnBi_2Te_4$ with 0.1g $MnCl_2$, temperature range was measured then all growth was prepared using small gradient temperature in the furnace. When the growth temperature is less than 550ºC then Mn-doped Bi_2Te_3 crystal is produced at the hot end and at the cold end Bi_2Te_3 liquid was noticed (see in Fig. 3(f)).

3. Extrinsic magnetic moments

Electron paramagnetic resonance (ESR) is used to investigate the extrinsic magnetic moments of Mn-doped Bi_2Ti_3 structure, where it takes the place of Bi ions. They obtained ferromagnetic polarization along a tetragonal axis after magnetic ordering and the orientation of Mn spins with a tetragonal axis due to tetragonal anisotropy. Note, the ESR spectrum structure of $Mn^{2+}(S=5/2)$ gets eliminated by ferromagnetic co-relation. In the presence of a large concentration (X > 0.02), the ESR signal of the Mn ion was found only in $Bi_{2-x}Mn_xTe_3$. ESR spectrum of the compound with x=0.05 has been discussed at a temperature range from 8 to 110K with sample orientation with respect to applied magnetic field H_o direction and parameters that strongly depend upon temperature [21]. The signal is present in the region between 3 to 4 KOe, in the ESR measurement when the parallel magnetic field is applied to the basal plane ab. A signal shifts towards a higher field on decreasing temperature. In $Bi_{1.95}Mn_{0.05}Te_3$, the magnetic ordering evidence of Mn ions is observed in vicinity of 10K (Fig. 4) by sharp strengthening of the resonance field H_R, and similar behavior of spectrum is founded for $Bi_{1.9}Mn_{0.09}Te_3$ with higher T_c of 12-13K [22,23]. Note, the resonance field H_R is transferred toward the lower field when a magnetic field is directed toward the hexagonal axis.

Superconducting quantum interference device (SQUID) magnetometry data indicates that the susceptibility (χ) can be fit to the curve of Weiss law:

$$\chi - \chi_0 = C(T - \theta)$$

Where θ, C and χ_o are the Weiss temperature, curie constant, and the temperature-independent term [24]. Upon cooling down to region 10K, integral intensity diverges and the inverse value is also shown in fig-4(b).

Resonance occurs due to susceptibility of spin, and similar behavior of ESR line integral intensity indicates that the line is produced from phase corresponding to dilute ferromagnetic $Bi_{1.95}Mn_{0.05}Te_3$. Let us discuss the resonant field for $Bi_{1.95}Mn_{0.05}Te_3$ that is predicted to become axial along a hexagonal axis and in the range of Curie temperature of 11K anisotropy field H_A emerges. Formula effective for the homogenous system is utilized to find a phenomenological approximation for resonance frequency and for external magnetic field situated in hexagonal plane $H_O>2H_A$, the resonance frequency is written as:

$$W_O^2 = \gamma^0 H_0(H_0 - 2H_A)$$

The resonance field value $H_0=H_r$ is transferred to a higher field for the fixed spectrometer because upon cooling down to 10K anisotropy field H_A is increased. The resonance frequency for external field directed alongside axis-c is given as

$$W = \gamma H_0 (H_0 - 2H_A)$$

According to this upon cooling the resonance field decreased (Fig-4(a), bottom) [24].

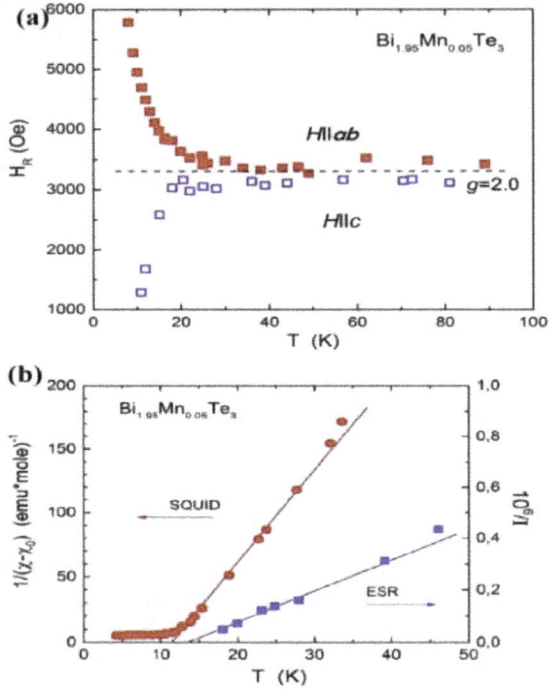

Figure 4. At two directions of external field, temperature-dependent resonant field is shown in Fig. 4(a) for ESR signal of $Bi_{1.95}Mn_{0.05}Te_3$, integral intensity and inverse magnetic susceptibility (χ) for $Bi_{1.95}Mn_{0.05}Te_3$ along temperature is represented in Fig. 4(b). Reused from [4].

4.   Intrinsic magnetic properties

Time-reversal symmetry in intrinsic topological insulators can break with magnetism with certain magnetic ordering, that joined together the other symmetry operation and time reversal operations like TC_n and T_t where C_n is the rotational operator, T is the time reversal operation and the translation operator is t. Z_2 topological properties are protected and extra constraints are provided by this equivalent time-reversal symmetry. For example, the bulk "axion angle" quantized to π due to the T_t symmetry operator in antiferromagnetic (AFM) topological insulator $MnBi_2Te_4$ (MBT) which is compulsory for making an axion insulator [25]. The Stoichiometric $MnBi_2Te_4$ septuple layer (SL) is formed as the result of Mn doping in Bi_2Te_4 instead of the chemical substitution of Bi by Mn. MBT heterostructure is formed due to SLs incorporation between Bi_2Te_4 quintuple layers and a 90 m eV topological gap was noticed in heterostructure which points out that in intrinsic topological insulators, quantum anomalous hall effect (QAHE) occurs at high temperatures. MBT holds ferromagnetic ordering within the septuple layers shown by theoretical measurement and the magnetic moment of the doped Mn atom is aligned opposite to the adjacent layer in the z-axis (out of the plane). For studying emergence of rich topological processes, intrinsic magnetic topological insulator $MnBi_2Te_4$ is an ideal system [26].

5.   Heterostructure comprising MBT and magnetic monolayer materials

For studying high-temperature quantum anomalous hall effect and stabilizing surface moment of MBT, 2D materials and construction of heterostructure from MBT is suggested. MBT-CrI_3 heterostructure characteristics were investigated by Fu et. al [27]. The magnetic ordering stability in the $MnBi_2Te_4$ surface layer was increased due to the CrI_3 monolayer ferromagnetic interaction with a surface layer of MBT. Large exchange bias (used to stabilize magnetization of soft ferromagnetic layer) of 40 m eV is produced by CrI_3 monolayer which is greater than the Neel temperature of MBT (2 m eV energy exists at 25K) as illustrated in Fig. 5(a). Band topology of MBT is slightly affected by CrI_3 monolayer, so we can say that the CrI_3 monolayer and MBT heterostructure help to study quantum anomalous hall effect temperature [26].

6.   MBT Family

Due to the ability to manifest the high-temperature QAHE, the $MnBi_2Te_4$ family has attracted the attention of researchers. For the investigation of non-trivial band structure and the magnetism of the $MnBi_2Te_4$ family, angle-resolved photoemission spectroscopy (ARPES) has been utilized. The surface band gap of Bi_2Te_4 is revealed in ARPES [28]. A huge family of Van der Waals materials have an intrinsic magnetic topological band, these materials are manifested as $MnBi_2Te_4(Bi_2Te_3)_n$ where n is 1,2,3.... as shown in Fig. 5. By the number n of Bi_2Te_3 layers, the topological magnetic properties are greatly adjustable because by increasing distance between MBT layers interlayer antiferromagnetic (AFM) interaction decreased. Ferromagnetic ground states are preferred then antiferromagnetic coupling for large values of n and interlayer antiferromagnetic coupling is maintained with Neel temperature 13K and 11K for small values of n like n=1 or 2 ($MnBi_4Te_7$ and

MnBi$_6$Te$_{10}$). Band calculation and ARPES experiment clarify that below Neel temperature along gapless side surface state MnBi$_4$Te$_7$ is an antiferromagnetic topological insulator [29]. The energy gap of MnBi$_4$Te$_7$ with Bi$_2$Te$_3$ termination is much greater compared to those with MBT termination and the bottom and top surface states are related to Bi$_2$Te$_3$ and MBT termination. Below the Neel temperature for n=2, MnBi$_6$Te$_{10}$ is an antiferromagnetic topological insulator possessing a full topological bulk gap and interestingly FM state is conserved and emerges when an applied magnetic field is lowered to zero and 0.1 T at 2K. 0.15 eV gap is present in the FM state and bulk ferromagnetic (FM) MnBi$_6$Te$_{10}$ is high order topological insulator similar to axion insulator alongside Θ=π and these calculations indicate that some MnBi$_6$Te$_{10}$ layers are intrinsic Chern insulator ([Bi$_2$Te$_3$]-[MnBi$_2$Te$_4$]-[Bi$_2$Te$_3$]-...) at zero magnetic fields with C = 1. Below 10.5 K, MnBi$_8$Te$_{13}$ lie into ferromagnetic phase showed in experimental transport result for n=3, APRES and First principles further illustrate MnBi$_8$Te$_{13}$ as intrinsic ferromagnetic axion insulator [26, 30].

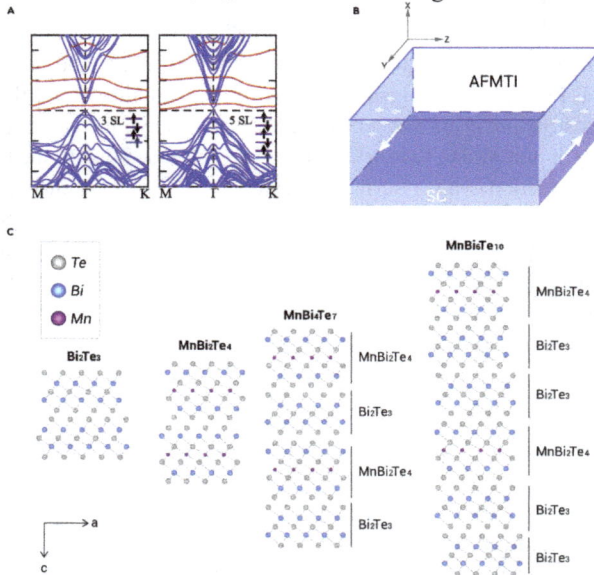

Figure. 5(a) One sextuple layer to six sextuple layer MBT band structure, in the insets Cr-eg band and MBT band are represented by red and blue lines. 5(b) arrows indicate Majorana modes and 1D chiral majorana edge states exhibited by hing of heterostructure. Structural characteristics and crystal structure of MBT illustrated in Fig. 5(c), MnBi$_2$Te$_4$ septuple layer is inserted as nBi$_2$Te$_3$ quintuple layers like n=1,2,3 for MBT, MnBi$_4$Te$_7$, and MnBi$_6$Te$_{10}$. Reused from [26].

When the intercalated Bi_2Te_3 layers are increased the result interaction between the nearest MBT septuple layers becomes too weak and MBT septuple layers are considered magnetically independent of each other below the critical temperature and magnetized along the z-axis. To obtain Chern insulator states, a small, large, or even zero external magnetic field is necessary because of fragile interlayer antiferromagnetic coupling and to understand topological superconductors. Furthermore, in $(MnBi_2Te_4)_2(Bi_2Te)$ heterostructure same axion insulator states are present and these states are easily analyzed in this MBT-related system that may make the path for a modern generation. In searching for new topological phases, MBT gives us a highly tunable and practical platform [31].

6.1   Chemically substituted MBT

One more process to tune the properties of $MnBi_2Te_4$ is a chemical substitution, by substituting Sb in place of Bi both magnetic and carrier properties can be adjustable, and by varying Sb atoms concentrations in $Mn(Sb_xBi_{1-x})_2Te_4$ topological effect can be controlled. The Neel temperature was reduced from 24K to 19K by varying the portion of Sb ($MnBi_2Te_4 \rightarrow MnSb_2Te_4$) reported by Yen et al. and for spin-flop transition critical magnetic field strength is necessary and moment saturation reduced dramatically [32]. All these conditions indicate that by increasing the Sb concentration x<0.86, antiferromagnetic coupling and anisotropy weaken. At x=0.63 in bulk $Mn(Sb_xB_{1-x})_2Te_4$ both p and n-type behaviors are also observed and by substitution of Sb, the Fermi level can be controlled. N-type conductivity of bulk material is mainly produced by high densities of Mn and Bi antisite defect and Sb substitution organized Fermi level and modified carrier concentration to bulk gap is the main hindrance in observing high temperature quantum anomalous effect. Making premium quality MBT single crystal is the second method to lower Fermi level. Experimentally $MnSb_2Te_4$ reported in both ways that it is antiferromagnetic (T_N – 19K) and ferromagnetic (T_C – 25K) material but theoretically reported only as antiferromagnetic material [31,33].

6.2   Puzzle surface state of MBT

Below Neel temperature ARPES experiment has manifested that surface states interchanging gap (10 to 200 m eV) is produced by the magnetic moment. However, it is very puzzling because the energy gap slightly changes with temperature, even a narrow band gap is observed above Neel temperature. So, we can say that an energy gap cannot arise from magnetic order. Conversely, some good resolution ARPES experiments predicted the bottom and top surface as having a gapless surface state, and this result shows conflict with the theoretical prediction for the surface state as shown in Fig. 6(a) [34]. Mn-3d state is ignored between energy ranges where non-trivial behavior arises and this state is located at 4 eV below the Fermi levels. So, according to Li et. al, it becomes hard to open the surface state exchange energy gap with magnetism. On the other hand, this observation opposes the Chern insulator phase and is supported/assisted by Density Functional Theory calculation. Bi_2Te_3/MBT heterostructure has been reported although at the Dirac point, there was a 90 m eV large magnetic gap and above Curie temperature structure vanished [35].

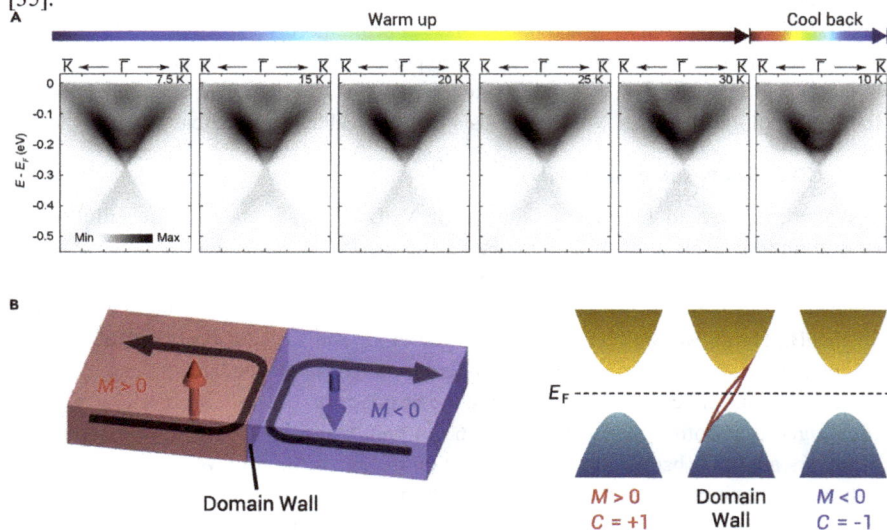

Figure 6. (a) Gapless surface at various temperatures investigated by ARPES. (b) Magnetic moment fluctuation model, a series of disorder domains formed by magnetic moment fluctuation and with different moments boundary states formed between irregular domains. Reused from [34].

This result represents that the interaction between topology in MBT and magnetism is conceivably strong to produce a magnetic gap. The gapless state can be described by spatially dependent magnetic moment manifested by Chen et. al [36]. Antiferromagnetic coupling between the underlying layer and surface layer is conceivably weaker than the antiferromagnetic coupling of bulk, which gives rise to the fluctuations of magnetic domains in the opposite and same direction. Gapless edge state interacting with energy gap present between domains when opposite sign gapped surface states produced by opposite magnetic moments. For the investigation of the gapless state, the electron density is provided by these edge states in ARPES experiment. A comparable explanation is also presented, which tells us the direction of surface magnetic domains and at the step edge of different domains topological states consequently occur. The surface gap is dissipated when the sample becomes paramagnetic by increasing temperature, this surface gap is reported at 50 m eV in a point contact experiment recently. A local magnetic gap exists in the MBT as shown by the results of point contact technology [26, 37].

7. Effect of magnetic moment on Mn atoms

Calculation with constraint functional should be done for investigation of Dirac point gap dependence on the magnetic moment value on the Mn atom, which gives sanctions until the difference between initial spin and desired spin is zero [38]. Alteration in the electronic structure of valence and conduction state of $MnBi_2Te_4$ and topological surface states is presented in Fig. 7(a1). As the initial value of the Dirac point gap was taken at 58 m eV (see Fig. 7(b) and Fig. 7(a3) represents the participation of localized states in 1st, $2^{nd,}$ and 3rd septuple layers manifested by blue and red signs. The magnetic moment reduces from 5.035 u_B to 4.38 u_B shown by the proposed dispersion map and the Dirac point reduces from 58 to 45 m eV. Fig. 7(b) indicates that gap size depends upon the change in a magnetic moment on the Mn atom, main point in this calculation is the linear approximation of zero gaps corresponding to non-zero magnetic moment on the Mn atom is generally twice less than its starting value. However, it is important to note in this calculation structure of the conduction band and valence band state does not change significantly in this system. When the magnetic moment is reduced, the electronic structure of the topological surface state also remains the same. Dirac point width change only and the valence band top remains in the same substitution region. When a magnetic Mn atom interchanges with a non-magnetic one, then due to substitution defect (Mn/Bi or MnTe) change in an effective magnetic moment for $MnBi_2Te_4$ is produced for short intervals. The opposite magnetic moment is observed when Mn is doped on the Te site which also results in a reduced exchange field acting on the topological surface state [39]. Fig. 7(c) shows that upon substitution 50% of Mn atom in the first septuple layer for non-magnetic Ge atom electronic structure and Dirac gap of topological surface states change. Dirac gap reduce to 11 m eV as a result of substitution and in designing synthetic layered topological structure Mn atom with a nonmagnetic one is useful [40].

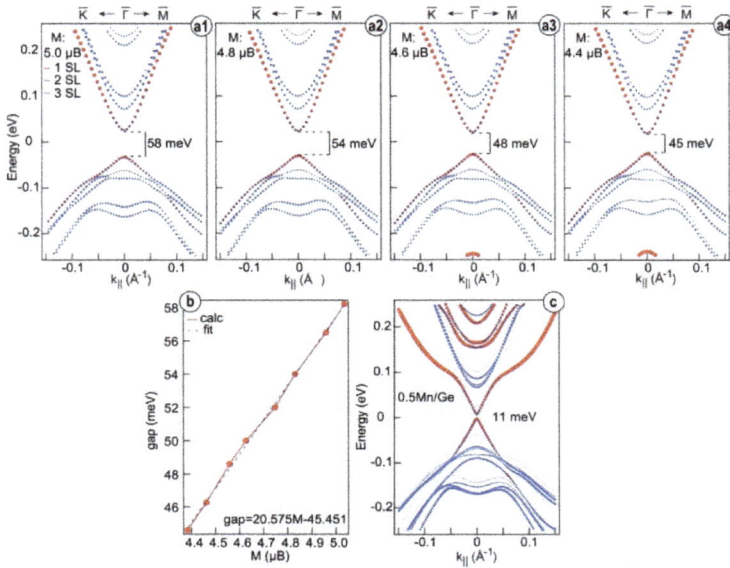

Figure 7. (a₁-a₄) demostrate Variation in topological surface states, electronic structure, closest conduction and valence band states for MnBi₂Te₄ with an indicated contribution of localized states in 1ˢᵗ (green symbol), 2ⁿᵈ/3ʳᵈ (blue symbol) in septuple layers upon a change in magnetic moment on Mn atom. (b) Dirac point energy gap depends upon magnetic moment on Mn atom. In the first septuple layer when 50% Mn atom was substituted in place of Ge atom then Dirac gap size was reduced and electron structure changed as shown in Fig. 7(c). Reused from [38].

8. Temperature evaluation of the electronic structure of MnBi₄Te₇

Temperature-dependent ARPES measurements across T_N were carried out to study the behavior of topological electronic structures in different magnetic phases as represented in Fig. 8. Topological surface states and dual temperature evaluation of the bulk state have been noticed across T_N as manifested in Fig. 8(a). Fig. 8(b) represented that the shape of dispersion does not change around Dirac plot from the zoomed-in intensity plot and Fig. 8(c) manifested that at Dirac point the energy distribution curve at Γ point retains a single peak structure and all these situations show that topological surface state with reducing gap remain in AFM phase transition [41].

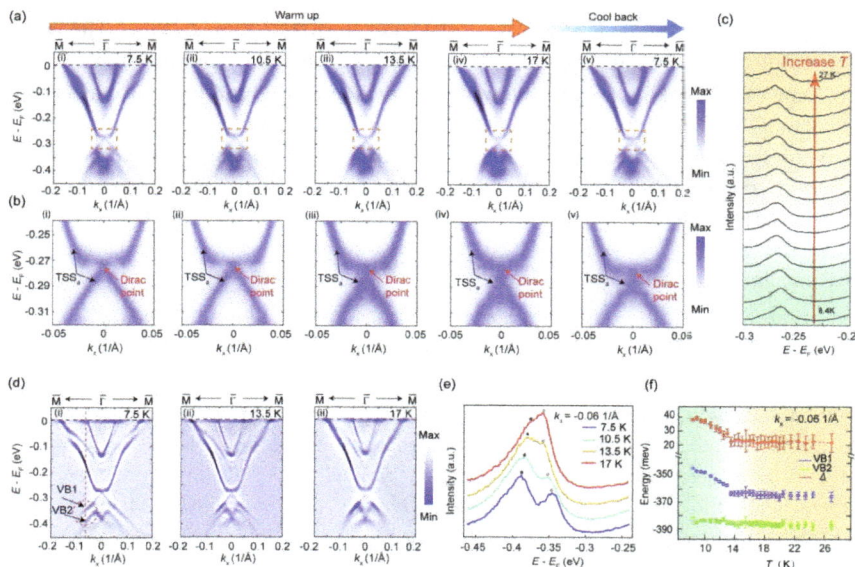

Figure 8. Temperature evaluation of the band structure of MnBi₄Te₇ as represented in Fig. 8(a) band dispression alongside M-Γ-M direction. Reduction in gap at specific temperature and near Dirac point zoom-in plot shown in Fig. 8(b), at Dirac point temperature evaluation by energy distribution curve (EDC) at Γ exhibit in Fig. 8(c) and 2ⁿᵈ derivative of photoemission intensity plot map comparison at (i) 7.5K (close to transition), (ii) 13.5K (antiferromagnetic phase) and (iii) 17K (paramagnetic phase) along M-Γ-M direction presented in Fig. 8(d). The most prominent variation is represented by arrows indicating VB2 and VB1. The location of energy distribution curve observed in Fig. 8(e, f) is designated by a dash line. Distribution energy curve stacked plot at Kₓ=0.006 Å from Fig. 8(a) and fitted peak positions indicated by empty and solid black circles corresponding to VB2 and VB1 manifested by Fig. 8(e and f). As function of temperature energy position of VB2 band, VB1 band, and energy difference between them is represented by green, blue, and red curves. Reused from [42].

In both CB and VB clear band evaluation has been found and by comparing photoemission intensity 2ⁿᵈ derivative plot calculated at 13K (close to transition) and at 7.5K (antiferromagnetic phase) the band structure difference is highlighted. The VB1 and VB2 band move toward each other clearly across paramagnetic and antiferromagnetic phase transition because of the relative shift between CB2 and topological surface states (TSSs) hybridization in CB2 and TSSs diminishes as illustrated in Fig. 8(d). The energy distribution curve peaks at various temperatures and different momenta are used to find the above-mentioned band's temperature evolution as shown in Fig. 8(e). As summarized in Fig. 8(e and f), the separation between VB2 and VB1 decreased to a minimum above T_N analyzed from extracted peak position. In MnBi$_4$Te$_7$ the strong correlation between magnetic ordering and electronic structure proposed by band variation over antiferromagnetic (AFM) transition. Due to Kz folding effect disappearance across antiferromagnetic–paramagnetic (AFM-PM) phase transition or interchange splitting from a ferromagnetic layer of MBT the reduction of splitting between VB2 and VB1 occurs [42].

9. Thermoelectricity in Mn doped topological insulator Bi$_2$Se$_3$

To construct future devices, we must be aware of the precise electrical domains inside lattice and electrons of outer orbit of magnetic doping agent in order to comprehend the electronic characteristics associated with magnetized doped topological insulators. It is difficult to preserve the balanced proportion, particularly in the most prevalent three-dimensional topological insulator, the tetra-dymites of form Bi$_2$Se$_3$. Making devices overuse in thermoelectric systems also requires optimizing carriers of charges. The way in which Mn-doped Bi$_2$Se$_3$ might be a viable option for a ferromagnetic topological insulator which is also appropriate for usage in thermoelectric devices [43].

9.1 Experimental setup

Highly pure elemental stoichiometric amounts of Te (99.999%), Bi (99.999%), and Mn (99.999%) were melted under a high vacuum in an evacuated quartz tube to form $Mn_xBi_{2-x}Se_3$ (where value of x is 0, 0.03, 0.05, and 0.1) single crystals. The combination was heated to 850°C for 48 hours, maintained there for another 24 hours, then gradually cooled to 650°C, where it was then annealed for another 24 hours before being quenched in cold water. They were split into shining single crystals. Tetradymite Bi$_2$Se$_3$ has a quintuple layer of Bi and Se that holds the crystal structure together. These VDW (Van der Waals) force-connected layers allow for the insertion of additional impurity between layers of atoms which leads to crystals being brittle and easy to break superconducting when Sr or Cu is included [43].

As seen in Fig. 9(a), an orthorhombic structure in the R − 3m space group is formed by the crystals for Mn concentration x= 0, 0.03, 0.05, and 0.1. This structure is characterized by dominating (001) peak. The cell parameters were determined by XRD data to be a = b = 4.15 Å, c = 28.69 Å, α = β = 90°, and Υ = 120° and the cell volume is 445.53a^3. With Mn doping, a substantial drop in the lattice parameter c was found. This is to be expected as the Mn ion's lower ionic radius than that of the Bi ion is an indirect confirmation of the dopant's successful substitution at the Bi sites. Fig. 9(b) displays the graph paper images of crystals. In Fig. 9(c), scanning electron microscope (SEM) images of a shattered corner is displayed around a 100 mm scale. At shattered corners, samples exhibit a layered shape. The EDX analysis spectrum of Mn-doped samples verify that bismuth and selenide are stoichiometrically balanced. Fig. 9(d) illustrates the high precision of single crystals.

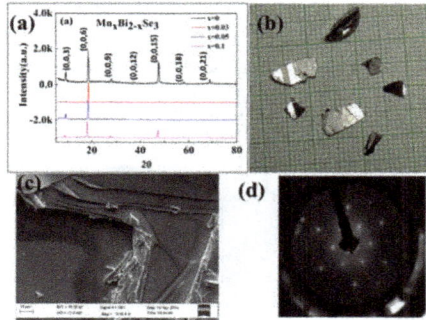

Figure 9. (a) $Mn_xBi_{2-x}Se_3$ XRD data peaks for concentration x=0, 0.03, 0.05 and 0.1. Crystal present on paper manifested in Fig. 9(b) and Fig. 9(c) illustrates scanning electron microscope picture of broken edges of crystal. (d) hexagonal lattic structure shown by Bi_2Se_3 crystal. Reused from [43].

9.2   Result and discussion

Fig. 10(a) displays the magnetic susceptibility (χ) for manganese doping with concentration $x = 0.03, 0.05, and\ 0.1$ in Bi_2Se_3 under the field of $1.5\ kOe$ to explore magnetic characteristics in the temperature range $5 - 300\ K$. The corresponding magnetic field has been placed to the measuring plane of the crystal. It has been found that the degree of sensitivity grows along with the amount of doping. The results were calculated using a modified version of the Curie-Weiss law provided by

$$\chi(T) = \frac{C}{T - \Theta_P} + \chi_0$$

where,

χ_0 = temperature independent susceptibility

Θ_P = Curie temperature and

C = Curie constant

The slope of adjusted lines in Fig. 10(a) shows such as for doping of concentration $x = 0.03$ obtained $\Theta_P \sim -7.9\,K$. The anti-ferromagnetic relationship of the Mn ion predominates at lower temperatures regardless of the diamagnetic character of the parent compound and is indicated by the negative value of the Curie temperature. Fig. 10(b) depicts an adiabatic magnetic curve at $2\,K$ for $x = 0.03$ and 0.1. There is no loop of hysteresis that can be observed for concentration x $= 0.03$ while for x $= 0.1$, a definite open hysteresis loop can be seen. It is found that the magnetic moment has a large magnitude and is still undifferentiated at $3\,T$ external field. A comparable study shows that 4% Mn doping of Bi_2Te_3 results in a second ordering phase shift with a ferromagnetic domain and $Tc = 9\,K$ [44].

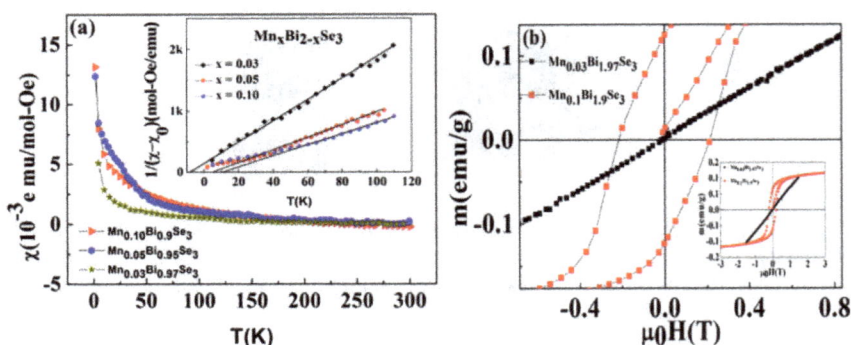

Figure 10. (a) Under zero field cooled protocol Mn doped Bi_2Se_3 magnetic susceptibility curve. Susceptibility data Curie-weiss fitting is illustrated in inset. Hysteries loop open at concentration x=0.1 and a linear curve of M-H is obtained for concentration x=0.03. Complete cycle of the loop is shown in inset. Reused from [43].

Conclusion

In summary, an investigation on $MnBi_2Se_4$ has been carried out in which it is observed that the surface is mostly terminated in Se-layer alongside to Van der Waal gap. For both samples n and p-type, Mn-doped Bi_2Se_3 is in high spin configuration proved by electron paramagnetic resonance experiment. Energy level $Mn^{1+}(d^6)$ is located outside the energy gap and energy level $Mn^{2+}(d^5)$ is present within valence band. In the fabrication of single crystal Mn doped Bi_2Se_3, it is identified that by increasing the Mn concentration the lattice parameter decreased and ferromagnetic co-relation exists for doping above 3%. The impact of intrinsic and extrinsic magnetic moment on topological insulators is different, this statement was made on the bases of data attained from SQUID magnetometry with ESR spectroscopy together. Even for a small amount of doping specific critical behavior was observed that confirm ferromagnetic ordering of Mn spin when magnetic Mn ion-doped to Bi_2Te_3.

Reference

[1] H. Zhang, C. X. Liu, X. L. Qi, X. Dai, Z. Fang, S. C. Zhang, Topological insulators in Bi_2Se_3, Bi_2Te_3, and Sb_2Te_3 with a single Dirac cone on the surface, Nat. Phys. 5 (2009) 438-442. https://doi.org/10.1038/nphys1270

[2] M. Z. Hasan, C. L. Kane, Colloquium: Topological insulators, Rev. Mod. Phys. 82 (2010) 3045. https://doi.org/10.1103/RevModPhys.82.3045

[3] A. Wolos, A. Drabinska, J. Borysiuk, K. Sobczak, M. Kaminska, A. Hruban, S. G. Strzelecka, A. Materna, M. Piersa, M. Romaniec, High-spin configuration of Mn in Bi_2Se_3 three-dimensional topological insulator, J. Magn. Magn. Mater. 419 (2016) 301-308. https://doi.org/10.1016/j.jmmm.2016.06.017

[4] V. Sakhin, E. Kukovitskii, N. Garifyanov, R. Khasanov, Y. Talanov, G. Teitelbaum, Local magnetic moments in the topological insulators, J. Magn. Magn. Mater. 459 (2018) 290-294. https://doi.org/10.1016/j.jmmm.2017.10.047

[5] V. Maurya, C. Dong, C. Chen, K. Asokan, S. Patnaik, High spin state driven magnetism and thermoelectricity in Mn doped topological insulator Bi_2Se_3, J. Magn. Magn. Mater. 456 (2018) 1-5. https://doi.org/10.1016/j.jmmm.2018.01.096

[6] M. Nakahara, Geometry, Topology and Physics, CRC press, 2003. https://doi.org/10.1201/9781420056945

[7] H. Zhang, C. X. Liu, X. L. Qi, X. Dai, Z. Fang, S. C. Zhang, Topological insulators in Bi_2Se_3, Bi_2Te_3, and Sb_2Te_3 with a single Dirac cone on the surface, Nat. Phys. 5 (2009) 438-442. https://doi.org/10.1038/nphys1270

[8] G. Checkelsky, J. Ye, Y. Onose, Y. Iwasa, Y. Tokura, Dirac-fermion-mediated ferromagnetism in a topological insulator, Nat. Phys. 8 (2012) 729-733. https://doi.org/10.1038/nphys2388

[9] H. W. Lee, B. S. Kim, S. Choi, J. Choi, J. Song, S. Cho, Mn-doped V_2VI_3 semiconductors: Single crystal growth and magnetic properties, J. Appl. Phys. 97 (2005) 10D324. https://doi.org/10.1063/1.1854451

[10] F. T. Huang, M. W. Chu, H. Kung, W. Lee, R. Sankar, S. C. Liou, K. Wu, Y. Kuo, F. Chou, Nonstoichiometric doping and Bi antisite defect in single crystal Bi_2Se_3, Phys. Rev. B 86 (2012) 081104. https://doi.org/10.1103/PhysRevB.86.081104

[11] P. Janicek, C. Drasar, P. Lostak, J. Vejpravova, V. Sechovsky, Transport, magnetic, optical and thermodynamic properties of $Bi_{2-x}Mn_xSe_3$ single crystals, Physica B: Condens. Matter 403 (2008) 3553-3558. https://doi.org/10.1016/j.physb.2008.05.025

[12] R. S. Silva, A. J. Gualdi, F. L. Zabotto, N. F. Cano, A. C. A. Silva, N. O. Dantas, Weak ferromagnetism in Mn^{2+} doped Bi_2Te_3 nanocrystals grown in glass matrix, J. Alloys Compd. 708 (2017) 619-622. https://doi.org/10.1016/j.jallcom.2017.03.066

[13] R. Silva, H. Mikhail, R. Pavani, N. Cano, A. Silva, N. Dantas, Synthesis of diluted magnetic semiconductor $Bi_{2-x}Mn_xTe_3$ nanocrystals in a host glass matrix, J. Alloys Compd. 648 (2015) 778-782. https://doi.org/10.1016/j.jallcom.2015.07.045

[14] T. Collins, Image J for microscopy, Biotechniques 43 (2007) S25-S30. https://doi.org/10.2144/000112517

[15] H. Zhang, W. Yang, Y. Wang, X. Xu, Tunable topological states in layered magnetic materials of $MnSb_2Te_4$, $MnBi_2Se_4$, and $MnSb_2Se_4$, Phys. Rev. B 103 (2021) 094433.

[16] S. Chowdhury, K. F. Garrity, F. Tavazza, Prediction of Weyl semimetal and antiferromagnetic topological insulator phases in Bi_2MnSe_4, NPJ Comput. Mater. 5 (2019) 33. https://doi.org/10.1038/s41524-019-0168-1

[17] R. C. Walko, T. Zhu, A. J. Bishop, R. K. Kawakami, J. A. Gupta, Scanning tunneling microscopy study of the antiferromagnetic topological insulator $MnBi_2Se_4$, Physica E, Low Dimens. Syst. Nanostruct. 143 (2022) 115391. https://doi.org/10.1016/j.physe.2022.115391

[18] C. Nowka, M. Gellesch, J. E. H. Borrero, S. Partzsch, C. Wuttke, F. Steckel, C. Hess, A. U. Wolter, L. T. C. Bohorquez, B. Buchner, Chemical vapor transport and characterization of $MnBi_2Se_4$, J. Cryst. Growth 459 (2017) 81-86. https://doi.org/10.1016/j.jcrysgro.2016.11.090

[19] J. Q. Yan, Z. Huang, W. Wu, A. F. May, Vapor transport growth of $MnBi_2Te_4$ and related compounds, J. Alloys Compd. 906 (2022) 164327. https://doi.org/10.1016/j.jallcom.2022.164327

[20] Y. Deng, Y. Yu, M. Z. Shi, Z. Guo, Z. Xu, J. Wang, X. H. Chen, Y. Zhang, Quantum anomalous Hall effect in intrinsic magnetic topological insulator $MnBi_2Te_4$, Science 367 (2020) 895-900. https://doi.org/10.1126/science.aax8156

[21] Y. Hor, P. Roushan, H. Beidenkopf, J. Seo, D. Qu, J. Checkelsky, L. Wray, D. Hsieh, Y. Xia, S.Y. Xu, Development of ferromagnetism in the doped topological insulator $Bi_{(2-x)}Mn_xTe_3$, Phys. Rev. B 81 (2010) 195203. https://doi.org/10.1103/PhysRevB.81.195203

[22] V. Sakhin, E. Kukovitskii, N. Garifyanov, Y. Talanov, G. Teitelbaum, Inhomogeneous state of the Bi_2Te_3 doped with manganese, J. Supercond. Nov. Magn. 30 (2017) 63-67. https://doi.org/10.1007/s10948-016-3801-y

[23] S. Zimmermann, F. Steckel, C. Hess, H. Ji, Y. S. Hor, R. J. Cava, B. Buchner, V. Kataev, Spin dynamics and magnetic interactions of Mn dopants in the topological insulator Bi_2Te_3, Phys. Rev. B 94 (2016) 125205. https://doi.org/10.1103/PhysRevB.94.125205

[24] V. Sakhin, E. Kukovitskii, N. Garifyanov, R. Khasanov, Y. Talanov, G. Teitelbaum, Local magnetic moments in the topological insulators, J. Magn. Magn. Mater. 459 (2018) 290-294. https://doi.org/10.1016/j.jmmm.2017.10.047

[25] J. Li, Y. Li, S. Du, Z. Wang, B. L. Gu, S. C. Zhang, K. He, W. Duan, Y. Xu, Intrinsic magnetic topological insulators in van der Waals layered $MnBi_2Te_4$-family materials, Sci. Adv. 5 (2019) eaaw5685. https://doi.org/10.1126/sciadv.aaw5685

[26] P. Wang, J. Ge, J. Li, Y. Liu, Y. Xu, J. Wang, Intrinsic magnetic topological insulators, The Innovation 2 (2021) 100098. https://doi.org/10.1016/j.xinn.2021.100098

[27] H. Fu, C. X. Liu, B. Yan, Exchange bias and quantum anomalous Hall effect in the $MnBi_2Te_4/CrI_3$ heterostructure, Sci. Adv. 6 (2020) 0948. https://doi.org/10.1126/sciadv.aaz0948

[28] Z. Jiang, J. Liu, Z. Liu, D. Shen, A review of angle-resolved photoemission spectroscopy study on topological magnetic material family of $MnBi_2Te_4$, Electron. Struct. (2022). https://doi.org/10.1088/2516-1075/acab47

[29] C. Hu, K. N. Gordon, P. Liu, J. Liu, X. Zhou, P. Hao, D. Narayan, E. Emmanouilidou, H. Sun, Y. Liu, A van der Waals antiferromagnetic topological insulator with weak interlayer magnetic coupling, Nat. Commun. 11 (2020) 97. https://doi.org/10.1038/s41467-019-13814-x

[30] C. Hu, L. Ding, K. N. Gordon, B. Ghosh, H. Li, S. W. Lian, A. G. Linn, H. J. Tien, C. Y. Huang, P. Reddy, Realization of an intrinsic, ferromagnetic axion insulator in $MnBi_8Te_{13}$, arXiv preprint arXiv.1910.(2019)12847

[31] W. Pinyuan, J. Ge, L. Jiaheng, L. Yanzhao, X. Yong, and W. Jian, Intrinsic magnetic topological insulators, The Innovation 2 (2021) 100098. https://doi.org/10.1016/j.xinn.2021.100098

[32] J. Q. Yan, S. Okamoto, M. A. McGuire, A. F. May, R. J. McQueeney, B. C. Sales, Evolution of structural, magnetic, and transport properties in $MnBi_{2-x}Sb_xTe_4$, Phys. Rev. B 100 (2019) 104409. https://doi.org/10.1103/PhysRevB.100.104409

[33] T. Murakami, Y. Nambu, T. Koretsune, G. Xiangyu, T. Yamamoto, C. M. Brown, H. Kageyama, Realization of interlayer ferromagnetic interaction in $MnSb_2Te_4$ toward the magnetic Weyl semimetal state, Phys. Rev. B 100 (2019) 195103. https://doi.org/10.1103/PhysRevB.100.195103

[34] B. Chen, F. Fei, D. Zhang, B. Zhang, W. Liu, S. Zhang, P. Wang, B. Wei, Y. Zhang, Z. Zuo, Intrinsic magnetic topological insulator phases in the Sb doped $MnBi_2Te_4$ bulks and thin flakes, Nat. Commun. 10 (2019) 4469. https://doi.org/10.1038/s41467-019-12485-y

[35] D. Rienks, L. Emile, W. Sebastian, J. S. Barriga, C. Ondrej, S. M. Partha, J. Ruzicka, N. Andreas, Large magnetic gap at the Dirac point in $Bi_2Te_3/MnBi_2Te_4$ heterostructures. Nature 576 (2019) 423-428. https://doi.org/10.1038/s41586-019-1826-7

[36] Y. J. Chen, L. X. Xu, J. H. Li, Y. W. Li, H. Y. Wang, C. F. Zhang, H. Li, Y. Wu, A. J. Liang, C. Chen, S. W. Jung, Topological electronic structure and its temperature evolution in antiferromagnetic topological insulator $MnBi_2Te_4$, Phys. Rev. X 9 (2019) 041040. https://doi.org/10.1103/PhysRevX.9.041040

[37] R. J. Hao, Y. Z. Liu, H. Wang, J. W. Luo, J. H. Li, H. Li, Y. Wu, Y. Xu, J. Wang, Detection of magnetic gap in topological surface states of $MnBi_2Te_4$, Chin. Phys. Lett. 38 (2021) 107404. https://doi.org/10.1088/0256-307X/38/10/107404

[38] A. Shikin, T. Makarova, A. Eryzhenkov, D. Y. Usachov, D. Estyunin, D. Glazkova, I. Klimovskikh, A. Rybkin, A. Tarasov, Routes for the topological surface state energy gap modulation in antiferromagnetic $MnBi_2Te_4$, Physica B: Conden. Matter 649 (2023) 414443. https://doi.org/10.1016/j.physb.2022.414443

[39] M. Garnica, M M. Otrokov, P.C. Aguilar, I.I. Klimovskikh, D. Estyunin, Z.S. Aliev, I. R. Amiraslanov, N.A. Abdullayev, V.N. Zverev, M. Babanly, Native point defects and their implications for the Dirac point gap at $MnBi_2Te_4$ (0001), NPJ Quantum Mater 7 (2022) 7. https://doi.org/10.1038/s41535-021-00414-6

Materials Research Forum LLC
https://doi.org/10.21741/9781644902851-6

[40] P. Kurz, F. Forster, L. Nordstrom, G. Bihlmayer, S. Blugel, Ab initio treatment of noncollinear magnets with the full-potential linearized augmented plane wave method, Phys. Rev. B 69 (2004) 024415. https://doi.org/10.1103/PhysRevB.69.024415

[41] D. Wortmann, P. Kurz, S. Heinze, K. Hirai, G. Bihlmayer, S. Blugel, Resolving noncollinear magnetism by spin-polarized scanning tunneling microscopy, J. Magn. Magn. Mater. 240 (2002) 57-63. https://doi.org/10.1016/S0304-8853(01)00733-7

[42] L. Xu, Y. Mao, H. Wang, J. Li, Y. Chen, Y. Xia, Y. Li, D. Pei, J. Zhang, H. Zheng, Persistent surface states with diminishing gap in $MnBi_2Te_4Bi_2Te_3$ superlattice antiferromagnetic topological insulator, Science Bulletin 65 (2020) 2086-2093. https://doi.org/10.1016/j.scib.2020.07.032

[43] V. Maurya, P. Neha, P. Srivastava, S. Patnaik, Superconductivity by Sr intercalation in the layered topological insulator Bi_2Se_3, Phys. Rev. B 92 (2015) 020506.

[44] J. S. Dyck, P. Hajek, P. Lostak, C. Uher, Diluted magnetic semiconductors based on $Sb_{2-x}V_xTe_3$ (0.01<~ x<~ 0.03), Phys. Rev. B 65 (2002) 115212.

Topological Insulators: Materials and Applications
Materials Research Foundations 154 (2024) 120-146

Materials Research Forum LLC
https://doi.org/10.21741/9781644902851-7

Chapter 7

Topological Insulators in Optical Applications

Sabahat Urossha[1] and S.S. Ali[1*]

[1]School of Physical Sciences, University of the Punjab, Lahore 54590, Pakistan

* shahbaz.sps@pu.edu.pk

Abstract

Since the discovery of topological insulators, researchers in condensed matter physics have focused on studying multiple topological states of matter. Topological insulators can transport an electron across the boundary without backscattering when they experience surface contaminants because of the unique exotic phase they exhibit. Research in topological photonics is among the most active fields of study in optics, and it is also one of the driving forces of research in topological physics. With the recent discovery of topological states of matter, EM waves can now be controlled and manipulated in a novel way. These metamaterials have the ability to revolutionize a wide range of electromagnetic design domains, from very durable cavities to tiny waveguides.

Keywords

Topological Insulators, Nonlinear Optical Behavior, Saturable Absorber

Abbreviations used

Photonic topological insulators	PTI
Topological insulators	TI
Photonic crystal	PC
Electromagnetic	EM
Photonic band gaps	PBG
Nonlinear Optical	NLO
Two-wave mixing	TWM
Beam splitter	BS
Spectroscopic ellipsometry	SE
Saturable absorbers	SA
Polarizer controller	PC
Amplified spectrum emission	ASE

Contents

1.   Introduction

Nontrivial unidirectional states of light can be supported by synthetic electromagnetic materials termed as photonic topological insulators [1]. EM wave counterparts of electronic topological phases explored in condensed matter physics are photonic topological phases. They can provide reliable unidirectional light propagation pathways, analogous to their electronic equivalents [2]. Unlike electronic topological insulators, photonic topological insulators (PTIs) are insulators in bulk and conductors upon the surface. Even in the presence of massive discontinuities, these structures sustain sustained edge modes that traverse without dissipation or backscattering [3-7].

Berkley researchers have demonstrated that photoelectrons emitted by topological insulators having spin-polarized electrons could be regulated through laser. Topological insulators (TIs) behave differently than other materials on Earth, and it's the electronic properties that make them so intriguing as metal-like properties are apparent on the surface of TIs, however, they are actually insulators in bulk form. The electrons on the surface are always spin-polarized because the spin of the electron is determined by its momentum and is always normal to the electron's traveling direction. TIs are a major topic in materials research because of their ability to maintain their characteristics even at room temperature. The spin polarization of photoelectrons (generated by the photoelectric effect) may be controlled by adjusting the polarization of the light emitted by a UV laser, according to a Nature Physics publication from a Berkeley-based team of scientists [8].

2. Light trapping in thin film

Energy generation remains one of the major issues in modern society, and it appears that utilizing renewable and clean energy sources, like solar energy, is the most effective approach. Solar cells must improve in efficiency and cost before they can be used as the world's primary source of energy. The primary cost factor for crystalline silicon solar cells is the silicon wafer at the moment. A major interest in developing thin film solar cells with absorber layer thicknesses of between several 100 nm and several microns has been expressed by research organizations worldwide. Solar cells, in contrast to this, reduce the absorption and short circuit current by reducing active layer thickness. Numerous new strategies for trapping and holding sunlight in cells' active layer have been developed by researchers in recent years to tackle this issue [9–24]

M.A. Shameli et al. suggested a new use for photonic TIs. They found that when photonic TIs are incorporated into a solar cell's active layer, light trapping can take place via the stimulation of edge states. As a result of violating the C6 symmetry, trivial topology is transformed into non-trivial topology, and edge modes are generated at the interface. Since those edge modes move sidewards in a solar cell, they have an increased probability of absorption and, as a result, an increased short circuit current occurs. They improved the concept by making use of multi-scale photonics TIs functioning at distinct wavelengths for the goal to be met with absorption enhancements into a broader area of wavelength. A full wave numerical analysis was used to review the performance of the suggested solar cell. The suggested idea worked well numerically and increased the cell's absorption throughout in a variety of wavelengths, as a result of 47 percent enhancement in the short circuit current of photovoltaic cells [25].

2.1 Solar cell embedded with photonic topological insulator

The structure suggested by *M.A. Shameli et al.* is shown in Fig. 1. As depicted in this figure, a photovoltaic cell incorporates along with photonic crystals based on two-dimensional PTIs to keep sunlight controlled and trapped in active layers via excited edge-states. The Ag backside interface of the solar cell has a thickness of $H_{Ag} = 100$ nm. The active layers of a photovoltaic cell are composed of Si with $H_{si} = 1500$ nm in height. N-type and P-type Si make up an active layer of designed photovoltaic cells. The p-type layer has a thickness of 1300 nm, whereas the n-type layer has a thickness of 200 nanometers. The p-type layer contains the PTIs, which are close to the p-n junction. The light would be concentrated in the p-n junction, here the created charge carriers have been divided and then transported towards the interface prior to recombination occurring because topological insulators prevent light from penetrating inside.

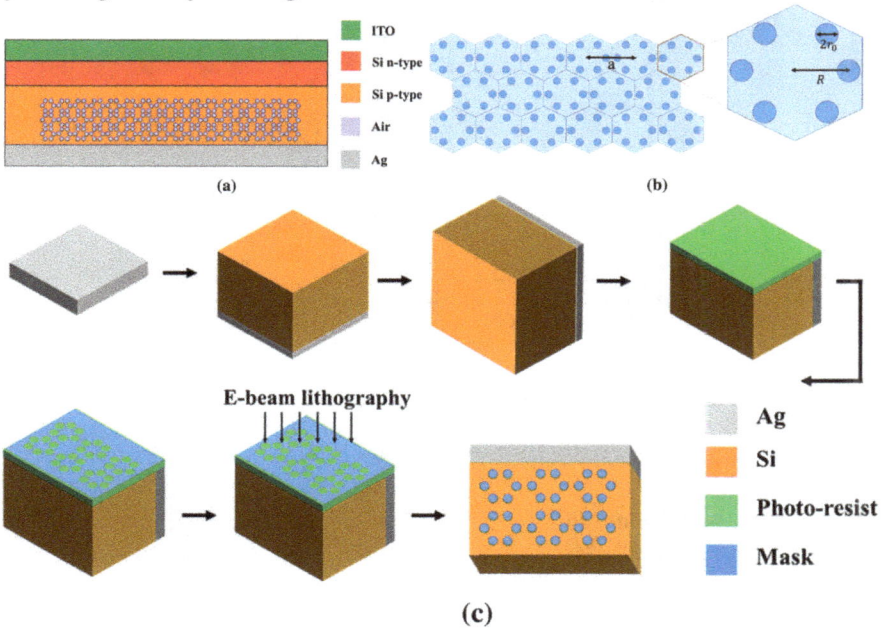

Figure 1. a) The PTIs is placed into thin-film solar cell. b) PTIs schematics, which are created using a photonic crystal with hexagonal arrangement. c) fabrication: firstly silver is coated with silicon layers. The photoresist layer is then put on structure afterwards the sample is rotated. The arrangement of PTIs is then shifted towards photo-resist layers using E-beam lithography. Lastly, the pattern is etched into silicon layer by the process of etching. Reprinted from [M. Ali Shameli and L. Yousef, Optics and Laser Technology 145 (2022) 107457] with the permission of Elsevier publishing.

The absorption of their suggested solar cell is estimated by making the use of a numerically computed electric field and depicted in Fig. 2. Due to the purpose of comparing, the absorption of a basic photovoltaic cell along with similar depth is computed that is depicted into the figure, illustrated in Fig. 2, At the majority of visible light wavelengths, suggested solar cell's absorption is greater than that of a normal solar cell [25].

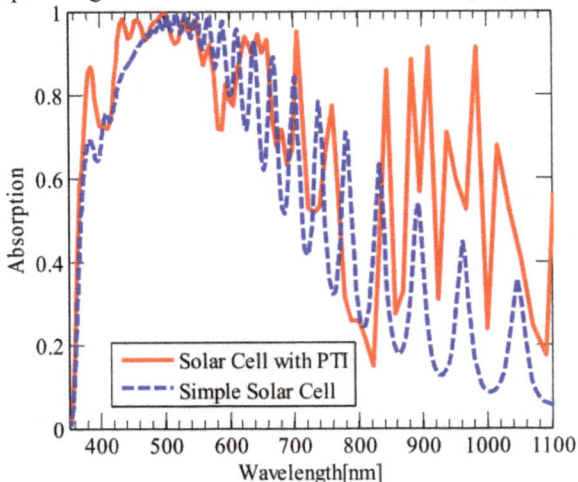

Figure 2.   The PTIs-based photovoltaic cell is examined within simple solar cell in terms of its absorption capacity. Reprinted from [M. Ali Shameli and L. Yousef, Optics and Laser Technology 145 (2022) 107457] with the permission of Elsevier publishing.

3.   Ultra wide dual bandwidth

The distinctive electromagnetic properties (EM) of artificial microstructure photonic crystals, such as their endurance and non-reciprocity, have been widely explored when they are periodically organized via mediums along with varying indices of refraction [26-32]. Topological insulators with a Chern number more than one can be created in ferromagnetic PCs, allowing for many unidirectional transmission states [33,34], but topological insulators, typically require extremely low temperatures or a powerful external magnetic field in order to operate. [35,36], due to why the photonic quantum Hall effect (QHE) is extremely difficult to achieve a sandwiched topological photonic crystal structure utilizing an E7 liquid crystal backdrop material and a C6 honeycomb medium column has been discovered by *Xiaofang Xu et al.* According to their findings, voltage can alter the operating bandwidth of a laminated photonic crystal using liquid crystal just like background material, and a laminated PC has a transmission efficiency of more than 90%. When merely a nontrivial layer of liquid crystal material is introduced, the working bandwidth of photonic crystal is altered. Due to its potential use in optical communication devices and Wavelength-Division Multiplexing (WDM) optical networks, the suggested TP crystals are extremely significant [37].

As depicted in Fig. 3 (a), when an exterior electric field is connected to 2 electrodes on the upper and bottom limits of topological PC to vary the nematicity of liquid crystal molecules, six silicon dielectric columns can be organized hexagonally in a regular pattern and can extend regularly. E7 liquid crystal is employed for the PC's background material. The crystal in a liquid has an index of refraction nbg 1.517 when no external voltage is supplied, but it increases to 1.741 when a particular voltage is given to 2 electrodes. Fig. 3 (b) It is a comprehensive depiction of C6 symmetry in $\overrightarrow{a1}$ and $\overrightarrow{a2}$ are unite lattice vectors. Fig. 3 (c) is a broad display of reddish lines in Fig. 1.3 (b). The coupling strength between the supercells can be changed by modifying the relative sizes of a and R, energy bands could be degenerating and inverted. The honeycomb lattice's first Brillouin zone is visible in the lower right corner.

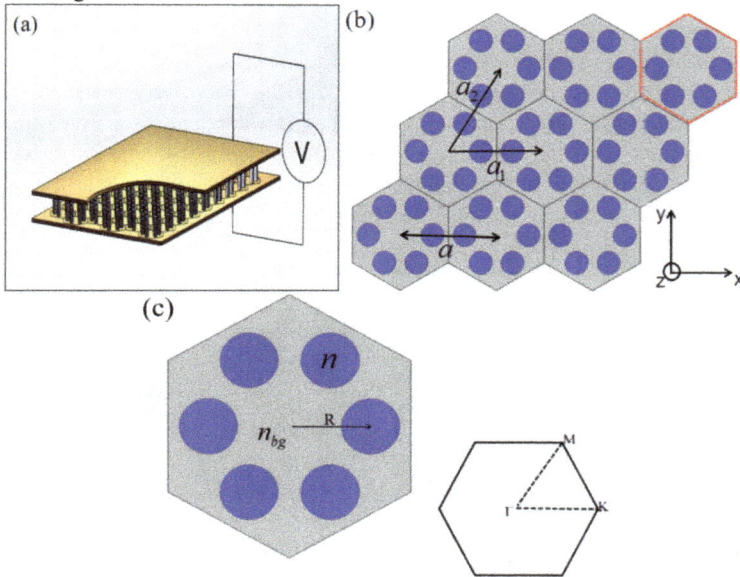

Figure 3.   Schematically of PC topology design. (a) A schematic depiction of a topological PC made up of si dielectric pillar. (b) In liquid crystal backgrounds (gray area), there are six silicon pillars (blue circles) on the left side. (c) a closer look of red dot hexagon in (b). Reprinted from [X. Xu et al., Results in Optics, 5 (2021) 100127] with the permission of Elsevier publishing [37].

In conclusion, it is depicted that liquid crystals in nontrivial structures cause a shift in the operating bandwidth of photonic crystals because of differences in refractive index between the nontrivial and trivial structures. Additionally, as an index of refraction of the non-trivial structure increases, the topological operating frequency of junction states drops over time. This topological PC structure might have been used in optical communication techniques like wavelength division multiplexing and beam splitter [37].

4. Topological beam splitter

A beam splitter is an essential component of advanced optical devices and optical systems that are both integrated and networked. Since the helical edge states have topological features, electromagnetic waves can easily pass through the structure's corners with no visible energy losses. A new topological beam splitter based on topological PCs has been presented by Yong-Feng Gao et al. Because of its basic structure, it is suited for miniaturization [38].

Recent research has shown that topological PCs could be constructed via deforming honeycomb lattices comprised entirely of dielectric material [39,40]. Another topological PC, this one made of elliptical silicon rods, has also been developed and tested at microwave frequencies [41]. Schematically Fig. 4 (a), is regarded as consisting of six elliptical rods that make up the unit cell presented in Fig. 4 (b), here ϵ_r and ϵ_o indicate dielectric constants, R is a distance among the cluster centers, and the dielectric cylinders. Parametric values illustrated previously are regarded as $\epsilon_r = 12$ and $\epsilon_o = 1$ and $a = 1$ μm.

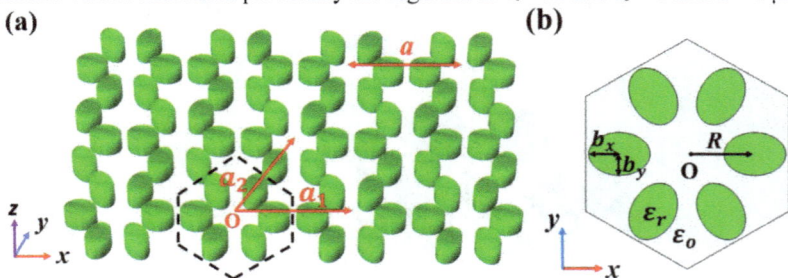

Figure 4. (a) An illustration of PCs with triangular lattice topology. The six elliptical rods (green) in each unit cell are formed of dielectric material (b) An elliptical rod has 2 major axes and minor axes, and their half-axis lengths are bx and by, respectively. Reprinted from [Y.-F. Gao, J.-P. Sun, N. Xu et al. Optics Communications 483 (2021) 126646] with the permission of Elsevier publishing [38].

4.1   Implementation of topological beam splitter

Yong-Feng Gao and co-workers were built using honeycomb lattice PCs, each with an elliptical rod of silicon placed in an air background, to create a topological beam splitter, as depicted in Fig. 5. The yellowish area indicates trivial PCs with $R = 0.303a$ and green region denotes non-trivial PCs along with $R = 0.350a$ likewise $b_x = 0.198a$ and $b_y = 0.09a$. This structure is chosen due to its small size and ease of one-way transmission. The system can be extremely stable if it is robust against flaws and backscattering is rigorously minimized. compared to PCs built of circular dielectric rods, in this configuration the light confinement is stronger, the PBG is wider, and spin guiding occurs when an Electromagnetic wave travels parallel to the interface [41].

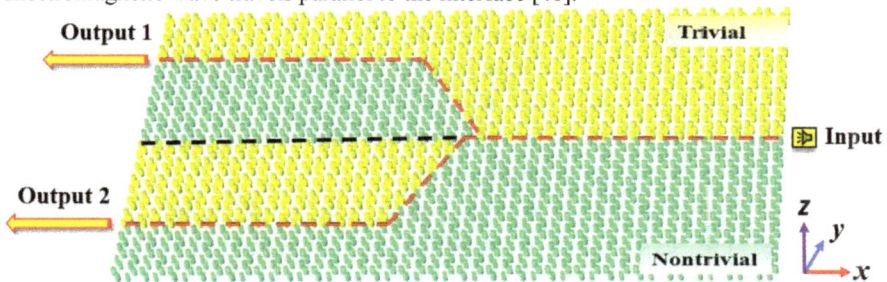

Figure 5.   Schematic of topological BS. Nontrivial lattice topological PCs form the green regions while trivial lattice PCs form the yellow. The trivial–nontrivial interfaces, shown as red dotted lines, can be considered the propagation route of an Electromagnetic wave in this setup. A block area is shown by the dotted black line. Reprinted from [Y.-F. Gao, J.-P. Sun, N. Xu et al. Optics Communications 483 (2021) 126646] with the permission of Elsevier publishing [38]

5.   Corner states in 2D photonic topological insulators

Edge states in TIs correlate to classical bulk-boundary correspondences: [42-44] n-dimensional (nD) TIs have (n-1)D gap-less edge states. According to a recent study into the field of higher-order topological insulators (HOTIs) that don follow this correlation, an nD HOTI brings about n-2D gap fewer edge-states, moreover to (n-1)D gapped edge states [45-55]. Wave manipulation is substantially facilitated by these lower-dimensional states, which open up new possibilities for photonic and integrated topological device development. *Mingxing Li et al.* created an optical C4 symmetric lattice structure that can generate 2 different kinds of corner states, the first of which was topological and the second was trivial. A new hierarchical response to dimensions was shown by merging topologically trivial and nontrivial PCs. This expands the study of higher-order photonic

TIs even further because it demonstrates a progression from bulk–corner–edge–corner–bulk. Mode area differences across corner states and their toughness is examined too [56].

Figure 6 represents their suggested PC on the basis of 2-dimensional square patterns. The simple unit is comprised of 8 dielectric cylinders along with a diameter of 3.6 millimeters that consist of si (relative permittivity $\varepsilon_d = 11.7$) and the air is underlying materials ($\varepsilon_r =$ one). $\overrightarrow{a1}$ and $\overrightarrow{a2}$ are 24 mm long unit vectors that represent the lattice constant 'a'. In determining the PC's band structure, there are two critical factors involving the closest distances d1 between the intra-cell cylinders and d2 among the inter-cell cylinders.

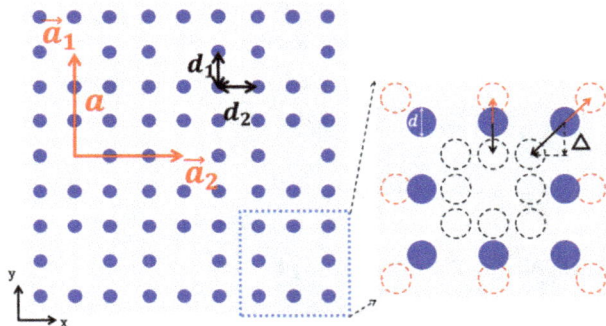

Figure 6. square lattice-based two-dimensional PC structure. The dashed blue represents basic unit. A basic unit is represented by blue dashes. The arrows in the right panel indicate the direction in which the cylinders are travelling in the magnified picture of a basic unit. The closest distances among intracell and intercell cylinders are d1 and d2, respectively. a is the lattice constant. Reprinted from [M. Li et al., J. Appl. Phys. 129, 063104 (2021)] with the permssion of AIP pubishing [56].

Topological PCs can be created by disrupting lattice symmetry, and the degenerate point of topological phase transition also has a significant impact in this process. *Mingxing Li et al.* first evaluate regardless of degenerate point can be found in the photonic band when Δ alters. Just transverse magnetic modes with electric fields that travel in the z direction are considered for the sake of simplicity. Figure 7 (a) depicts there is a direct correlation between energy band frequencies and distance covered Δ at 1st Brillouin zone border, and also when Δ = 0 the 3rd and 4th bands will degenerate.. Figure 7 (b) depicts bandgap development when Δ varies from negative to positive, and mostly simulations to produce findings conducted by Comsol Multiphysics.

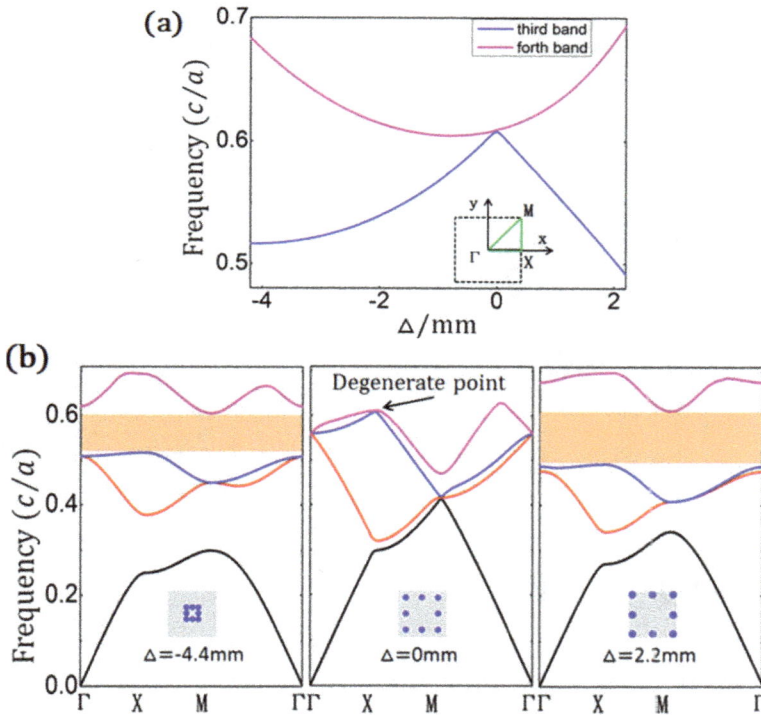

Figure 7. (a) Changing parameter Δ changes the trend of 3rd and 4th band at the X-point (b) Band structure from left to right, Δ = - 4.4 mm, 0 and 2.2 mm. This structure's whole bandgap may be seen in the orange area; the inset shows a unit of it. Reprinted from [M. Li et al., J. Appl. Phys. 129, 063104 (2021)] with the permission of AIP pubishing [56].

The relation between band-gap size and factor Δ has been studied, in order to learn more about the trend of bandgap variation, shown in Fig. 8 (a). By altering Δ, In terms of bandgap evolution, there are three types of PCs. The blue area indicated the first type. When Δ < - 2.85 mm, the size of the photonic bandgap increases as the value of Δ lowers, as does the fact that the bandgap appears to be complete, depicted black line in blue area of Fig. 8(a).

Connecting two PCs has resulted in the creation of a new design, the projected band is depicted in Fig. 8(b). The red curve indicates that an interface state exists between the two bulk states (gray area). For the purpose of verifying the edge states, a full-wave simulation was run, and the frequency of origin is adjusted to 0.552(c/a), which relates to the boundary state [blue dots].

The electric field (E_z) is depicted in Fig. 8 (c), and the fields are discovered to be fixed at the two structures' interface and degrade rapidly far apart from the interface.

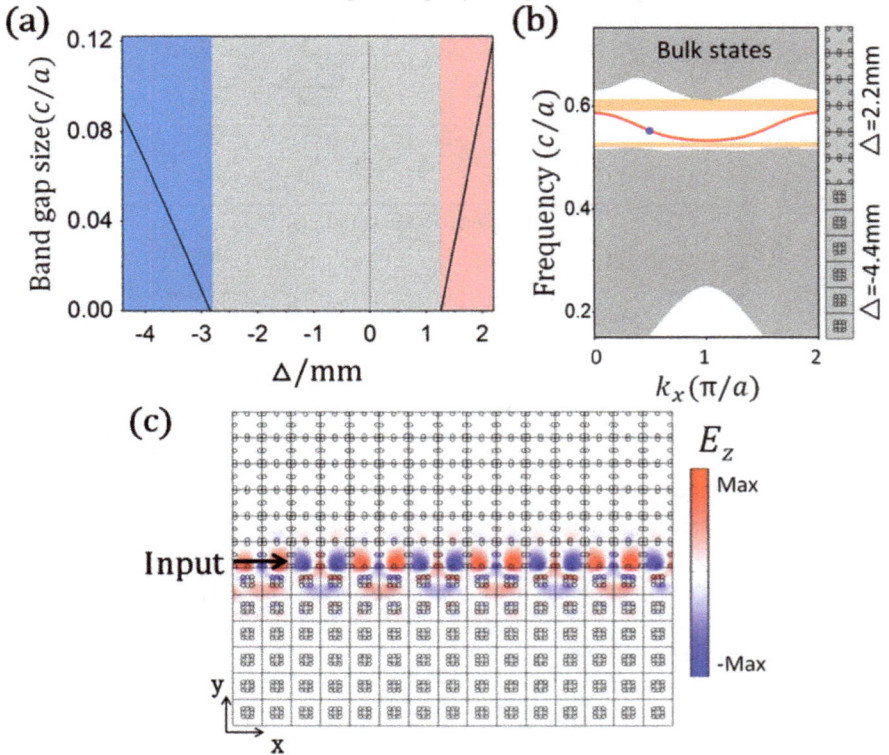

Figure 8   (a) The graph below shows link among geometric parameter Δ and bandgap. A whole bandgap exists in PC, including a topological trivial (blue area) and nontrivial (red area). (b) The projected band of 2 PCs in various topological phases, with red slope representing edge state and orange representing bandgap. The composite construction is on right side. (c) (E_z) of edge state alongwith frequency = 0.552(c/a) [blue dots into figure (b)]. Reprinted from [M. Li et al., J. Appl. Phys. 129, 063104 (2021)] with the permssion of AIP pubishing [56].

6. Bi₂Te₃ topological insulators

Bi_2Te_3 is an interesting nanometric 3D TIs alloy whose optical characteristics have attracted considerable attention [57]. whose bandgap fluctuation caused by photo-induced atomic displacement serves a critical part in TIs implementations [58] They can be used as semiconductors because of their shielded conductivity and optical properties [59]. Because of QCE which could alter the bandgap, electron's DOS, and Seebeck coefficient, Bi_2Te_3 is an effective and desirable thermoelectric (TE) material too [60]. The use of Bi_2Te_3 in the thermoelectric conversion process appears intriguing [61]. Bi_2Te_3 is essential for a wide range of TE implementations, including TE generators [62] higher-efficiency TE devices [63], heating, and cooling [64]. In recently investigations two-photon absorption (TPA) has been found to be extremely strong in a multilayer Bi_2Te_3 TI. [65]. The broadband non-linear absorption and broad nonlinear refraction of Bi_2Te_3 were examined too. Bi_2Te_3 was also examined for its broadband nonlinear absorption and large nonlinear refraction. Topological insulator-based photonics may employ linear and non-linear optical (NLO) responses of Bi_2Te_3 to create optoelectronic devices [66], optical limiting techniques [67], and PD devices [68].

6.1 Photo-induced structured waves

E.A. Hurtado-Aviles et al. proposed that Bi_2Te_3 nanostructures can be manufactured using a low-cost chemical method. As topological insulators, Bi2Te3 nanostructures are the primary focus of their research. Non-linear optical properties of Bi2Te3 suggest that it may be the suitable material for use in ultrafast uses and photonic instruments. The NLO response of Bi_2Te_3 thin films was studied using the two-wave mixing (TWM) method. Experimentally and numerically, the effect of a refractive index based on irradiance was studied. Fig. 9 shows a schematic depiction of the experimental set-up used in the study. Briefly, 532 nm "λ" of laser light system was adjusted through the use of focallength lens (LH) of 750 mm. The 4 ns and 100 mJ optical pulsation was generated using a linear polarised laser beam. After passing via a beam splitter (BS), the optical signal is divided in 2 coherent plane waves. By positioning symmetrical mirrors (M1-M2) at silmilar distance, pump and probe beams of varied intensity are focused upon sample. The pump beam's polarization was controlled via an achromatic half-wave plate (WP) and monitored with a photodetector (PD1) [69].

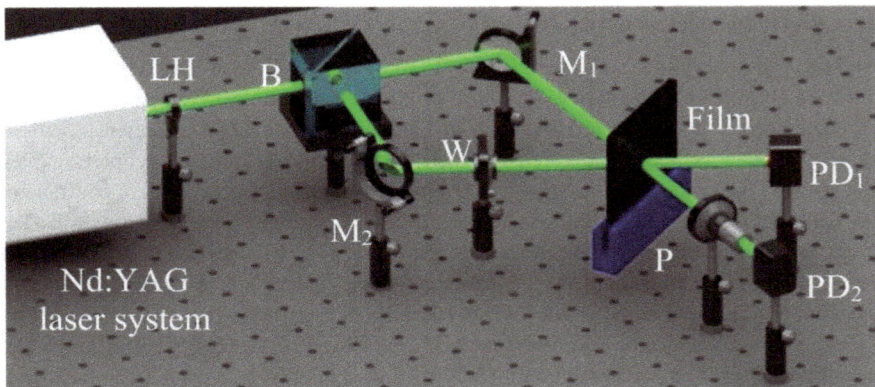

Figure 9. *TWM in the Bi_2Te_3 nanostructured sample was measured using an experimental setup. Reprinted from [E.A. Hurtado-Aviles et al. Optics and Laser Technology 140 (2021) 107015] with the permission of Elsevier publishing.*

Fig. 10(a) depicts a multi-layer optical model used to calculate the thickness of the Bi_2Te_3 sample in film form using spectroscopic ellipsometry (SE). A filter was used for the 2D typical SEM micrograph ascertained through their investigation to create the 3D graphic illustration. At the surface of the Bi_2Te_3 film, the 3D image shows a grainy and rather rough appearance. An optical model of the examined thin film was created using the DeltaPsi2® software and used to help explain the SE results. Nanostructure-integrated Bi2Te3 nanostructures have a rough layer thickness of L_2 in the TiO_2 film. At an incidence angle of 70°, the optical properties were measured and analyzed. Before and after UV irradiation, the test was performed in the presence of a sample or not exhibiting a photo-chromic effect Fig. 10(b) presents the experimentation and adequate design data for the index of refraction n, and extinction coefficient k [69].

The conclusion of this work shows that photoinduced energy transfer can be significantly affected by nonlinear optical phenomena that are polarization selective. Bi_2Te_3 has an optical band gap of 1.44 electron Volt, and IR-active mode A_{1u}^2 at 120.2 cm^{-1} in Raman spectrum both confirm that it is a topological insulator. The Bi_2Te_3 nanostructures were found to exhibit capacitive impedance characteristics via EIS testing. The distinct characteristics of the Bi_2Te_3 thin film and 3rd-order Non-linear optical effects can be taken into consideration for the development of photonic platforms and nonlinear systems [69].

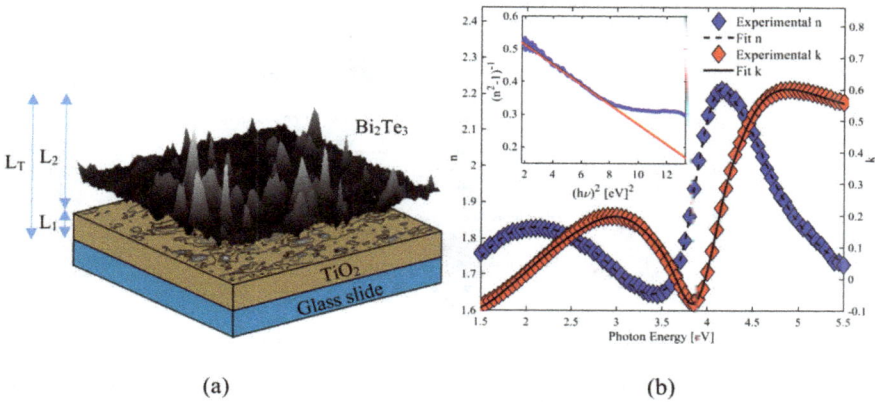

Figure 10. *(a) Realistic optical model was used for SE interpretations of Bi₂Te₃ film. (b) SE calculations and values of adequately design of nano-structured Bi₂Te₃ film, inset refers relation of refractive index factor $(n^2 - 1)^{-1}$ vs $(hv)^2$. Reprinted from [E.A. Hurtado-Aviles et al. Optics and Laser Technology 140 (2021) 107015] with the permission of Elsevier publishing.*

6.2   Dynamic optical study

S. Nimanpure et al. reported PCA based upon Terahertz experimental setup for Terahertz-TDS (Terahertz-Time domain spectroscopy). Researchers have demonstrated the growth of a Bi₂Te₃ TI thin film upon recently generated Carbon nanotubes with many walls on a versatile paper (MWCNT-FP) that were employed as the substrate for great response time in terms of delay and absorption into the Terahertz region for flexible THz electronics. THz-TDS within topological properties in (2D) bulk TI was also reported by the authors. Non-linear electro-dynamics and electro-optic behavior of Bi₂Te₃ TIs thin film produced on MWCNT-FP have been investigated as novel characteristics. It was also investigated whether the Bi₂Te₃ topological insulator thin film had a non-linear Terahertz transmittance. Dirac electron responses on the film's surface are thought to be responsible for this non-linear Terahetz behavior.

To illustrate the interaction of Bi₂Te₃ TIs with THz incident pulses, a schematic model are bases on MWCNT-FP (TI CNT hybrid) is presented in Fig. 11.

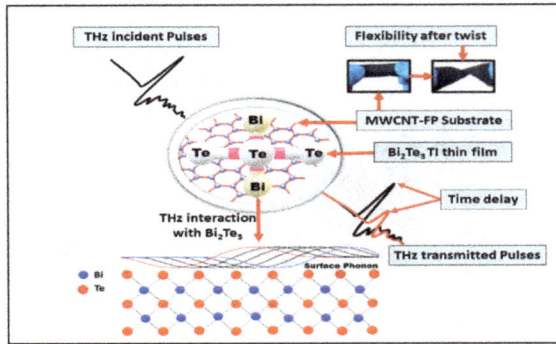

Figure 11.   *The interaction of incident Terahertz pulses alongwith a Bi_2Te_3 topological insulator basis upon MWCNT-FP thin film is depicted schematically. Incident THz pulse induces surface phonons. Reprinted from [S. Nimanpure et al., Optical Materials 121 (2021) 111490] with the permission of Elsevier publishing.*

The optical density and absorption coefficient are examined over a frequency range of 0.02–3.0 Terahertz. Fig. 1.12(a), depicts the absorption increases linearly in the frequency range of 0.02–1.0 Terahertz, In the range of 1.0–2.0 THz, the amplitude of absorption progressively increases and then gradually decreases. Free-carrier absorption might allow for a wide range of absorption, and afterward 1.0 Terahertz. Fig. 1.12(b), displays the optical density spectra of Bi_2Te_3 Topological insulators basis upon MWCNT-FP utilizing powerful Terahertz pulses at room temperature. The optical density rises as the frequency rises. Around 1.0 THz, the highest peak may be found. Up to 1.0 THz, the optical density spectrum moves upward; afterward 2.0 Terahertz, the optical density spectrum moves downward. Nonlinear signals can be clearly seen in this type of behavior. Due to its high amplitude, TI CNT hybrid thin film is responsible for the sample's non-linear behavior [70].

THz conductivity measurements of Bi_2Te_3 TPIs thin film upon MWCNT-FP were made at room temperature as depicted in Fig. 1.13. There are two conductivities in the TI CNT hybrid thin film: one for the real conductance and one for the imaginary conductance. THz conductivity is also studied like conductance from complicated transmission spectrum, with the same pattern. The σ_r and σ_i are directly related to real and imaginary conductance respectively having inverse relation with the thickness of the samples.

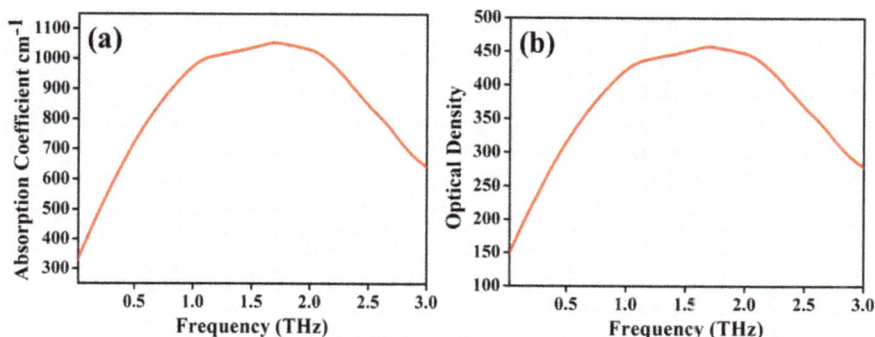

Figure 12. (a) Measurements of a Bi_2Te_3 TIs thin film sample yielded an absorption coefficient at THz frequencies. (b) At THz frequencies, the optical density of the observed samples. The logarithm scale of the absorption coefficient is used to calculate the optical density. Reprinted from [S. Nimanpure et al., Optical Materials 121 (2021) 111490] with the permission of Elsevier publishing.

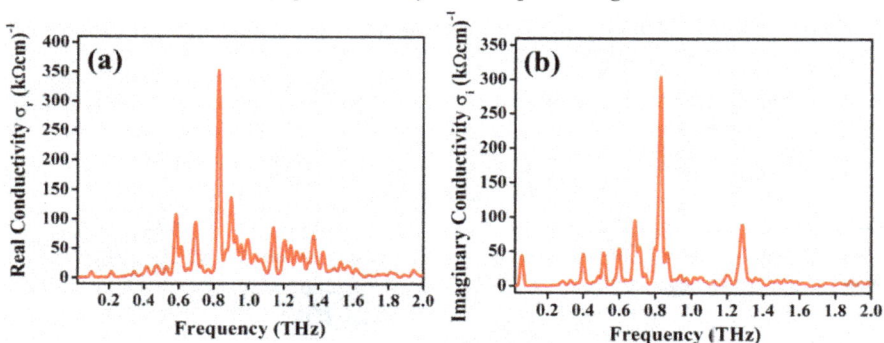

Figure 13. (a) The real σ_r representing the main characteristis at lower frequency (0.02–0.8 THz) and highest frequency (> 0.8 THz). At low frequency end, the inter subband transition appears, while phonons appear at the maximum frequency end. The σ_i is depicted in (b). Reprinted from [S. Nimanpure et al., Optical Materials 121 (2021) 111490] with the permission of Elsevier publishing.

7. Bi_2Se_3 topological insulator

The short band-gap bulk and the gapless surface enable Topological insulators to have a wide bandwidth of SA. Bismuth Selenide (Bi_2Te_3) has significantly lower saturation intensity [71] and this Bi_2Te_3 could be significantly beneficial in developing lower threshold pulses laser [72–74]. Bi_2Te_3 as Q-switcher has been illustrated in [75-78]. Chen et al. [76] reported pulse repetition rates ranging from 4.5 kilohertz to 12.9 kilohertz, pulse widths ranging from 13.4 µs to 36 µs, and pulsation energies ranging from 11.8 µJ to 13 µJ with Q-switched EDFL. Luo et al. [77] showed a 1 µm Q-switched fiber into a linearized cavity within a 7–29 kHz repetition rate, 2–8 µs pulse width, and 5–16 nJ pulse energy. In contrast to this, Luo et al. [75] illustrated 2 µm QYDF along with a recurrence interval of 8.4-26.8 kHz , pulsation depth ranging of 4-18 microsecond, and pulses energy ranging from 0.1-0.3 µJ was carried out. Chen et al. [76] illustrated that the Q-switched EDFL could be adjusted to operate at wavelengths ranging from 1510.9 to 1589.1 nanometers, within recurrence rates ranging from 2 to 12.

7.1   Saturabe absorber

H. Haris et al. illustrated Q-switched (EDFL) and Ytterbium-doped fiber laser (YDFL) based on Bi2Se3-like Saturable absorber. The recurrence of the pulse could be set between 14.9 kHz and 62.5 kHz in a 1.5 nm area. The repeated rate exhibited in this experiment is far higher than that of the previous studies [75–78]. In order to prove that the manufactured Bi_2Se_3 is a wideband Saturable absorber, reliable Q-switched lasers may be generated in both the 1.0 and 1.5 µm ranges.

Fig. 14(a) depicts the experimental setup for suggested Q-switched EDFLA 1480 nm laser diode pumps in the cavity, whose output is coupled to a 1 m (EDF) gain medium. The core and cladding diameters of the EDF employed in this experiment are 4 and 125 micrometers, respectively. TI Bi_2Se_3 SA (TI Bi2Se3 – SA) placed upon fiber ferrule is coupled to another fiber ferrule. At 1550 nm, the insertion loss of Bi_2Te_3– SA is 1.5 dB. The laser cavity retains 95% of light, while the output taps out 5% in order to take a reading. The isolator is put into the cavity to make sure that the oscillating laser is unidirectional. The output is linked to a 50/50 coupler so that an optical spectrum analyzer (OSA) along a spectral resolution of 0.02 nanometer can measure amplified spectrum emission (ASE) and a 1.2 Gigahertz bandwidth PD can monitor the Q-switched pulses at the same time the experiment was reperformed with YDF instead of EDF. Q-switched Ytterbium Doped Fiber experimental setup is depicted in Fig. 14(b). The YDF employed is 1 meter long. At 1100 nm, the SA's insertion loss was measured to be 1.4 dB [79]. To summarize, a robust EDQF and YDQF laser using a Topological Insulator Bi_2Te_3–Saturable absorber has been developed. Topological insulator Bi_2Te_3–SA has a saturation intensity of 90.2 MW//cm². In combination with TI Bi_2Se_3-SA, the proposed EDFL and YDFL generate stable pulsed Q-switching voltages. This shows that the TI Bi_2Te_3– SA that was built is indeed a wide-bandwidth saturable absorber device.

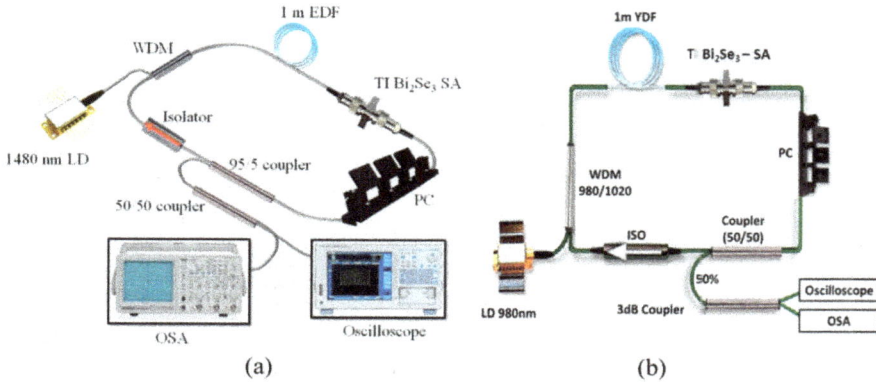

Figure 14. (a) Experimental design for suggested Q-switched EDFL (b) and for Q-switched YDFL. Reprinted from [H. Haris et al, Optics & Laser Technology 88 (2017) 121–127] with the permission of Elsevier publishing.

Y. Cheng et al. presented results at 604 nm utilizing saturable (SAs) absorbers made from a topological insulator (TI) Bi_2Te_3 nanosheet laser diode-pumped Pr:LiYF4 laser. On its peak of 130kHz, the device can provide pulses with an average power of 26 (mW) and an energy roughly 0.2μJ . Bi_2Te_3 saturable absorbers as a result of their study now have a working wavelength in the visible range.

Fig. 15 depicts the experimental laser setup. Laser diodes of the type InGaN are utilized as pump sources up to a maximum power output of roughly 2W along with a wavelength of about 444 nm. To rectify the pump beam's astigmatism, the laser diode is merged with a lens having f = 3 mm. This 5-millimeter longer and 0.5% doped gain material received a pump beam after being focused by the 50-mm (focal length) lens. To protect the laser crystal, an indium surface wrap was used, and a copper block was used to house it. Pulsation thickness is reduced by around 1050–802 nanoseconds (see Fig. 16), resulting in a single energy of up to 0.2 μJ with 26 mW output power [80]. The study performed by Y. Cheng et al. indicated that optimizing the thermal effect and transfer quality of the TI Bi_2Te_3 thin film can lead to further improvements in Q-switching laser performance [80].

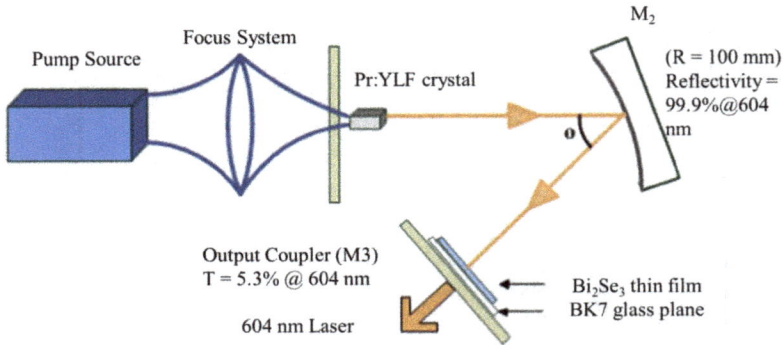

Figure 15. *Schematically experiment setup of Bi_2Te_3 bases uopn Q-switched Pr:YLF laser at 604 nanometer. Reprinted from [Y. Cheng et al., Optics & Laser Technology 88 (2017) 275–279] with the permission of Elsevier publishing.*

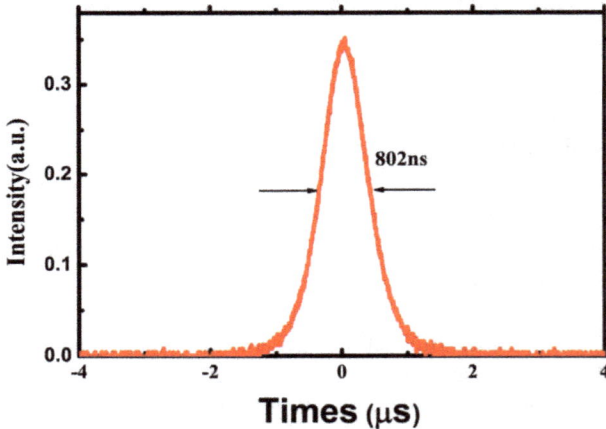

Figure 16. *With a pump power of 1.357 W, the narrowest pulse width was obtained. Reprinted from [Y. Cheng et al., Optics & Laser Technology 88 (2017) 275–279] with the permission of Elsevier publishing.*

Conclusions

In this chapter, we have examined only a few aspects of topological insulators, with an emphasis on optical properties and applications. Numerous excellent experimental findings have proved the revolutionary power of topological photonics for both fundamental and applied photonics. More realistic revolutionary approaches aim to redesign laser systems by providing topological toughness and quantum-limited traveling wave amplifiers that are safe from both internal loss and backscattering. In the future, topological photonics have the potential to move beyond the linear regime and into the realm of strong photon–photon and photon–matter interactions. It is also possible to build an effective multi-physics platform on polaritons, which are topological polaritonic excitations that arise from strong light–matter coupling, such as in a 2-Dimensional quantum system.

References

[1] L. Lu, D. John, Topological photonics, Nature Photonics 8 (2014) 821-829. https://doi.org/10.1038/nphoton.2014.248

[2] T. Ozawa, H. Price, A. Amo, N. Goldman, M. Hafezi, L. Lu, M. Rechtsman, C. Mikael, D. Schuster, J. Simon, O. Zilberberg, I. Carusotto, Topological photonics, Reviews of Modern Physics 91 (2019) 015006. https://doi.org/10.1103/RevModPhys.91.015006

[3] M.A. Shameli, L. Yousef, Absorption enhanced thin-flm solar cells using fractal nano-structures, IET Optoelectron. (2021). https://doi.org/10.1049/ote2.12036

[4] L.H. Wu, X. Hu, Scheme for achieving a topological photonic crystal by using a dielectric material, Phys. Rev. Lett. 114 (2015), 223901. https://doi.org/10.1103/PhysRevLett.114.223901

[5] M.H. Latifpour, L. Yousef, Topological plasmonic edge states in a planar array of metallic nanoparticles, Nanophotonics 8 (2019) 799-806. https://doi.org/10.1515/nanoph-2018-0230

[6] L. Lu, J.D. Joannopoulos, M. Solja˘ci'c, Topological photonics, Nat. Photonics 8 (2014) 821-829. https://doi.org/10.1038/nphoton.2014.248

[7] P. Zhou, G.G. Liu, X. Ren, Y. Yang, H. Xue, L. Bi, L. Deng, Y. Chong, B. Zhang, Photonic amorphous topological insulator, Light Sci. Appl. 9 (2020) 1-8. https://doi.org/10.1038/s41377-020-00368-7

[8] C.H. Park. C. Hwang, Photoelectron spin-flipping and texture manipulation in a topological insulator, Nature Physics 9 (2013) 293-298. https://doi.org/10.1038/nphys2572

[9] W. Ye, R. Long, H. Huang, Y. Xiong, Plasmonic nanostructures in solar energy conversion, J. Mater. Chem. 5 (2017) 1008-1021. https://doi.org/10.1039/C6TC04847A

[10] H.A. Atwater, A. Polman, Plasmonics for improved photovoltaic devices, Nat. Mater 9 (2010) 205-213. https://doi.org/10.1038/nmat2629

[11] J. Jang, M. Kim, Y. Kim, K. Kim, S.J. Baik, H. Lee, J.C. Lee, Three-dimensional a-Si: H thin-film solar cells with silver nano-rod back electrodes, Curr. Appl Phys. 14 (2014) 637-640. https://doi.org/10.1016/j.cap.2014.02.006

[12] P. Yu, Y. Yao, J. Wu, X. Niu, A.L. Rogach, Z. Wang, Effects of plasmonic metal core dielectric shell nanoparticles on the broadband light absorption enhancement in thin film solar cells, Sci. Rep. 7 (2017) 1-10. https://doi.org/10.1038/s41598-016-0028-x

[13] Y.M. Song, J.S. Yu, Y.T. Lee, Antireflective submicrometric gratings on thin-film silicon solar cells for light-absorption enhancement, Opt. Lett. 35 (2010) 276-278. https://doi.org/10.1364/OL.35.000276

[14] F. Taghian, V. Ahmadi, L. Yousef, Enhanced thin solar cells using optical nanoantenna induced hybrid plasmonic traveling-wave, J. Lightwave Technol. 34 (2016) 1267-1273. https://doi.org/10.1109/JLT.2015.2511542

[15] M.A. Shameli, L. Yousef, Absorption enhancement in thin-film solar cells using an integrated meta surface lens, JOSA B 35 (2018) 223-230. https://doi.org/10.1364/JOSAB.35.000223

[16] M.R. Khan, X. Wang, P. Bermel, M.A. Alam, Enhanced light trapping in solar cells with a meta-mirror following generalized Snell's law, Opt. Express 22 (2014) A973-A985. https://doi.org/10.1364/OE.22.00A973

[17] M.A. Shameli, P. Salami, L. Yousef, Light trapping in thin film solar cells using a polarization independent phase gradient metasurface, J. Opt. 20 (2018) 125004. https://doi.org/10.1088/2040-8986/aaea54

[18] W.R. Erwin, H.F. Zarick, E.M. Talbert, R. Bardhan, Light trapping in mesoporous solar cells with plasmonic nanostructures, Energy Environ. Sci. 9 (2016) 1577-1601. https://doi.org/10.1039/C5EE03847B

[19] L.H. Zhu, M.R. Shao, R.W. Peng, R.H. Fan, X.R. Huang, M. Wang, Broadband absorption and efficiency enhancement of an ultra-thin silicon solar cell with a plasmonic fractal, Opt. Express 21 (2013) A313-A323. https://doi.org/10.1364/OE.21.00A313

[20] P. Kowalczewski, M. Liscidini, L.C. Andreani, L. Claudio Andreani, Engineering Gaussian disorder at rough interfaces for light trapping in thin-film solar cells, Opt. Lett. 37 (2012) 4868. https://doi.org/10.1364/OL.37.004868

Materials Research Forum LLC
https://doi.org/10.21741/9781644902851-7

[21] D.H. Lee, J.Y. Kwon, S. Maldonado, A. Tuteja, A. Boukai, Extreme light absorption by multiple plasmonic layers on upgraded metallurgical-grade silicon solar cells, Nano Lett. 14 (2014) 1961-1967. https://doi.org/10.1021/nl4048064

[22] S. Liu, R. Jiang, P. You, X. Zhu, J. Wang, F. Yan, Au/Ag core-shell nanocuboids for high-efficiency organic solar cells with broadband plasmonic enhancement, Energy Environ. Sci. 9 (2016) 898-905. https://doi.org/10.1039/C5EE03779D

[23] M.H. Muhammad, M.F.O. Hameed, S.S.A. Obayya, Broadband absorption enhancement in modified grating thin-film solar cell, IEEE Photonics J. 9 (2017) 1-14. https://doi.org/10.1109/JPHOT.2017.2698720

[24] M.H. Mohammadi, D. Fathi, M. Eskandari, Nio@GeSe core-shell nano-rod array as a new hole transfer layer in perovskite solar cells: A numerical study, Sol. Energy 204 (2020) 200-207. https://doi.org/10.1016/j.solener.2020.04.038

[25] M.A. Shameli, L. Yousef, Light trapping in thin-film crystalline silicon solar cells using multi-scale photonic topological insulators, Optics & Laser Technology 145 (2022) 107457 https://doi.org/10.1016/j.optlastec.2021.107457

[26] K.V. Klitzing, G. Dorda, M. Pepper, New method for high-accuracy determination of the fine-structure constant based on quantized hall resistance, Phys. Rev. Lett. 45 (1980) 494-497 https://doi.org/10.1103/PhysRevLett.45.494

[27] Y. Zhang, Y. Tan, H.L. Stormer, P. Kim, Experimental observation of the quantum Hall effect and Berry's phase in graphene, Nature 438 (2005) 201-204 https://doi.org/10.1038/nature04235

[28] K.S. Novoselov, Z. Jiang, Y. Zhang, Room-temperature quantum hall effect in graphene, Science 315(2007) 1379. https://doi.org/10.1126/science.1137201

[29] X. Cheng, C. Jouvaud, X. Ni, S.H. Mousavi, A.Z. Genack, A.B. Khanikaev, Robust reconfigurable electromagnetic pathways within a photonic topological insulator, Nat. Mater. 15 (2016) 542-548 https://doi.org/10.1038/nmat4573

[30] M. Goryachev, M.E. Tobar, Reconfigurable microwave photonic topological insulator, Phys. Rev. Appl. 6 (2016) 064006. https://doi.org/10.1103/PhysRevApplied.6.064006

[31] A. Furusaki, Weyl points and Dirac lines protected by multiple screw rotations, Sci. Bull. 62 (2017) 788-794 https://doi.org/10.1016/j.scib.2017.05.014

[32] A. Khaleque, H.T. Hattori, Absorption enhancement in graphene photonic crystal structures, Appl. Opt. 55 (2016) 2936-2942 https://doi.org/10.1364/AO.55.002936

[33] S.A. Skirlo, L. Lu, M. Soljacic, M., Multimode one-way waveguides of large Chern numbers, Phys. Rev. Lett. 113 (2014) 113904. https://doi.org/10.1103/PhysRevLett.113.113904

[34] S.A. Skirlo, L. Lu, Q. Yan, Experimental observation of large chern numbers in photonic crystals, Phys. Rev. Lett. 115 (2015) 253901. https://doi.org/10.1103/PhysRevLett.115.253901

[35] C.L. Kane, E.J. Mele, Z2 Topological order and the quantum spin hall effect, Phys. Rev. Lett. 95 (2005) 146802 https://doi.org/10.1103/PhysRevLett.95.226801

[36] D.N. Sheng, Z.Y. Weng, L. Sheng, F.D.M. Haldane, Quantum spin hall effect and topologically invariant Chern numbers. Phys. Rev. Lett. 97 (2006) 036808 https://doi.org/10.1103/PhysRevLett.97.036808

[37] X. Xu, X. Guo, S. Mu, H. Zhang, J. Huang, A topological photonic crystal with ultra-wide dual bandwidth, Results in Optics 5 (2021) 100127 https://doi.org/10.1016/j.rio.2021.100127

[38] Y. Gao, J. Sun, N. Xu, Z. Jiang, Q. Hou, H. Song, M. Jin, C. Zhang, Manipulation of a topological beam splitter based on honeycomb photonic crystals, Optics Communications 483 (2021) 126646 https://doi.org/10.1016/j.optcom.2020.126646

[39] L. Wu, X. Hu, Scheme for Achieving a Topological Photonic Crystal by Using Dielectric Material, Phys. Rev. Lett. 114 (2015) 223901. https://doi.org/10.1103/PhysRevLett.114.223901

[40] X. Zhu, H. Wang, C. Xu, Topological transitions in continuously deformed photonic crystals, Phys. Rev. B 97 (2017) 085148 https://doi.org/10.1103/PhysRevB.97.085148

[41] Z. Jiang, Y. Gao, L. He, et al., Phys. Chem. Chem. Phys. 21 (2019) 11367. https://doi.org/10.1039/C9CP00789J

[42] M. Z. Hasan and C. L. Kane, Colloquium: Topological insulators, Rev. Mod. Phys. 82 (2010) 3045. https://doi.org/10.1103/RevModPhys.82.3045

[43] D. Xiao, M. C. Chang, and Q. Niu, Berry phase effects on electronic properties, Rev. Mod. Phys. 82 (2010) 1959. https://doi.org/10.1103/RevModPhys.82.1959

[44] T. L. Hughes, R. G. Leigh, and O. Parrikar, Torsional anomalies, Hall viscosity, and bulk-boundary correspondence in topological states, Phys. Rev. D 88 (2013) 025040. https://doi.org/10.1103/PhysRevD.88.025040

[45] B. Y. Xie, G. X. Su, H. F. Wang, H. Su, X. P. Shen, P. Zhan, M. H. Lu, Z. L. Wang, and Y. F. Chen, Visualization of higher-order topological insulating phases in two-

dimensional dielectric photonic crystals, Phys. Rev. Lett. 122 (2019) 233903. https://doi.org/10.1103/PhysRevLett.122.233903

[46] X. D. Chen, W. M. Deng, F. L. Shi, F. L. Zhao, M. Chen, and J. W. Dong, Direct observation of corner states in second-order topological photonic crystal slabs, Phys. Rev. Lett. 122 (2019) 233902. https://doi.org/10.1103/PhysRevLett.122.233902

[47] B. Y. Xie, G. X. Su, H. F. Wang, F. Liu, L. Hu, S.-Y. Yu, P. Zhan, M.-H. Lu, Z. Wang, and Y. F. Chen, Higher-order quantum spin Hall effect in a photonic crystal, Nat. Commun. 11 (2020) 3768. https://doi.org/10.1038/s41467-020-17593-8

[48] W. A. Benalcazar, B. A. Bernevig, and T. L. Hughes, Quantized electric multipole insulators, Science 357 (2017) 61. https://doi.org/10.1126/science.aah6442

[49] C. W. Peterson, W. A. Benalcazar, T. L. Hughes, and G. Bahl, A quantized microwave quadrupole insulator with topologically protected corner states, Nature 555 (2018) 346. https://doi.org/10.1038/nature25777

[50] Z. D. Song, Z. Fang, and C. Fang, (d−2)-dimensional edge states of rotation symmetry protected topological states, Phys. Rev. Lett. 119 (2017) 246402.

[51] M. Geier, L. Trifunovic, M. Hoskam, and P. W. Brouwer, Second-order topological insulators and superconductors with an order-two crystalline symmetry, Phys. Rev. B 97 (2018) 205135. https://doi.org/10.1103/PhysRevB.97.205135

[52] M. Ezawa, Higher-order topological insulators and semimetals on the breathing Kagome and pyrochlore lattices, Phys. Rev. Lett. 120 (2018) 026801. https://doi.org/10.1103/PhysRevLett.120.026801

[53] H. D. Xue, Y. H. Yang, F. Gao, Y. D. Chong, and B. L. Zhang, Acoustic higher-order topological insulator on a kagome lattice, Nat. Mater. 18, (2019) 108. https://doi.org/10.1038/s41563-018-0251-x

[54] M.S. Garcia, V. Peri, R. Susstrunk, O. R. Bilal, T. Larsen, L. G. Villanueva, and S. D. Huber, Observation of a phononic quadrupole topological insulator, Nature 555 (2018) 342. https://doi.org/10.1038/nature25156

[55] X. J. Zhang, H. X. Wang, Z. K. Lin, Y. Tian, B. Y. Xie, M. H. Lu, Y. F. Chen, and J. H. Jiang, Second-order topology and multidimensional topological transitions in sonic crystals, Nat. Phys. 15 (2019) 582. https://doi.org/10.1038/s41567-019-0472-1

[56] M. Li, Y. Wang, M. Lu, and T. Sang, Two types of corner states in two-dimensional photonic topological insulators, J. Appl. Phys. 129 (2021) 063104. https://doi.org/10.1063/5.0039586

[57] M. Shaik, I.A. Motaleb, Investigation of the optical properties of PLD-grown Bi2Te3 and Sb2Te3, IEEE Int. Conf. Electro/Inf. Technol. (2013) 1-6.

[58] M. Hada, K. Norimatsu, S. Tanaka, S. Keskin, T. Tsuruta, K. Igarashi, Bandgap modulation in photoexcited topological insulator Bi2Te3 via atomic displacements, J. Chem- Phys. 145 (2016) 024504. https://doi.org/10.1063/1.4955188

[59] G. Hao, X. Qi, Y. Liu, Z. Huang, H. Li, K. Huang, Ambipolar charge injection and transport of few-layer topological insulator Bi2Te3 and Bi2Se3 nanoplates, J. Appl. Phys. Am. Inst. Phys. AIP (2012) 114312. https://doi.org/10.1063/1.4729011

[60] S. Yazdani, M.T. Pettes, Nanoscale self-assembly of thermoelectric materials: A review of chemistry-based approaches, Nanotechnology 29 (2018). https://doi.org/10.1088/1361-6528/aad673

[61] M.T. Pettes, J. Maassen, I. Jo, M.S. Lundstrom, L. Shi, Effects of surface band bending and scattering on thermoelectric transport in suspended bismuth telluride nanoplates, Nano Lett. 13 (2013) 5316-5322. https://doi.org/10.1021/nl402828s

[62] H. Goldsmid, Bismuth telluride, and its alloys as materials for thermoelectric generation, Materials (Basel). 7 (2014) 2577-2592. https://doi.org/10.3390/ma7042577

[63] I. Bejenari, V. Kantser, Thermoelectric properties of bismuth telluride nanowires in the constant relaxation-time approximation, Phys. Rev. B - Condens. Matter Mater. Phys. 78 (2008) 115322. https://doi.org/10.1103/PhysRevB.78.115322

[64] I.T. Witting, T.C. Chasapis, F. Ricci, M. Peters, N.A. Heinz, G. Hautier, The thermoelectric properties of bismuth telluride, Adv. Electron. Mater. 5 (2019) 1800904. https://doi.org/10.1002/aelm.201800904

[65] J. Qiao, M.Y. Chuang, J.C. Lan, Y.Y. Lin, W.H. Sung, R. Fan, Two-photon absorption within layered Bi2Te3 topological insulators and the role of nonlinear transmittance therein, J. Mater. Chem. C. 7 (2019) 7027-7034. https://doi.org/10.1039/C9TC01885A

[66] L. Miao, J. Yi, Q. Wang, D. Feng, H. He, S. Lu, Broadband third order nonlinear optical responses of bismuth telluride nanosheets, Opt. Mater. Express. 6 (2016) 2244. https://doi.org/10.1364/OME.6.002244

[67] M. Molli, S. Parola, L.A. Avinash Chunduri, S. Aditha, V. Sai Muthukumar, T. Mimani Rattan, Solvothermal synthesis and study of nonlinear optical properties of nanocrystalline thallium doped bismuth telluride, J. Solid State Chem. (2012) 85-89. https://doi.org/10.1016/j.jssc.2011.11.051

[68] J.L. Liu, H. Wang, X. Li, H. Chen, Z.K. Zhang, W.W. Pan, High performance visible photodetectors based on thin two-dimensional Bi2Te3 nanoplates, J. Alloy. Compd. 25 (2019) 656-664. https://doi.org/10.1016/j.jallcom.2019.05.299

[69] E.A. Aviles, M. Valdez, J. A. Torres, C.J. Torres, H. Guti'errez , C. Torres, Photo-induced structured waves by nanostructured topological insulator Bi2Te3, Optics & Laser Technology 140 (2021) 107015. https://doi.org/10.1016/j.optlastec.2021.107015

[70] M. Jewariya, M. Nagai, K. Tanaka, Ladder climbing on the anharmonic intermolecular potential in an amino acid microcrystal via an intense monocycle terahertz pulse, Phys. Rev. Lett. 105 (2010) 203003. https://doi.org/10.1103/PhysRevLett.105.203003

[71] Z. Luo, Y. Huang, J. Weng, H. Cheng, Z. Lin, B. Xu, Z. Cai, H. Xu, 1.06 μm Q switched ytterbium-doped fiber laser using few-layer topological insulator Bi2Se3 as a saturable absorber, Opt. Express 21 (2013) 29516-29522. https://doi.org/10.1364/OE.21.029516

[72] Z. Luo, C. Liu, Y. Huang, D. Wu, J. Wu, H. Xu, Z. Cai, Z. Lin, L. Sun, J. Weng, Topological-insulator passively Q-switched double-clad fiber laser at 2 μm wavelength, IEEE J. Sel. Top. Quantum Electron. 20 (5) (2014) 1-8. https://doi.org/10.1109/JSTQE.2014.2305834

[73] C. Yu, Z. Chujun, H. Huihui, C. Shuqing, T. Pinghua, W. Zhiteng, L. Shunbin, Z. Han, W. Shuangchun, T. Dingyuan, Self-assembled topological insulator: Bi2Se3 the membrane as a passive q-switcher in an erbium-doped fiber laser, J. Lightwave Technol. 31 (17) (2013) 2857-2863. https://doi.org/10.1109/JLT.2013.2273493

[74] Y. Chen, C. Zhao, H. Huang, S. Chen, P. Tang, Z. Wang, Self-assembled topological insulator: Bi Se membrane as a passive Q-switcher in an erbium-doped fiber laser, J. Lightwave Technol. 31 (2013) 2857-2863. https://doi.org/10.1109/JLT.2013.2273493

[75] Z. Luo, Y. Huang, J. Weng, H. Cheng, Z. Lin, B. Xu, 1.06 μm Q-switched ytterbium-doped fiber laser using few-layer topological insulator Bi2Se3 as a saturable absorber, Opt. Express 21 (2013) 29516-29522. https://doi.org/10.1364/OE.21.029516

[76] Y. Chen, C. Zhao, S. Chen, J. Du, P. Tang, G. Jiang, Large energy, wavelength widely tunable, topological insulator Q-switched erbium-doped fiber laser, IEEE J. Sel. Top. Quantum Electron. 20 (2014) 315-322. https://doi.org/10.1109/JSTQE.2013.2295196

[77] Z. Luo, C. Liu, Y. Huang, D. Wu, J. Wu, H. Xu, Topological-insulator passively Q-switched double-clad fiber laser at 2 μm wavelength, IEEE J. Sel. Top. Quantum Electron. 20 (2014) 1-8. https://doi.org/10.1109/JSTQE.2014.2305834

[78] W. Richter, H. Kohler, C.R. Becke, A Raman and far-infrared investigation of phonons in the rhombohedral V2-VI3 compounds bismuth tritelluride Bi2Te3, bismuth triselenide Bi2Se3, antimony tritelluride Sb2Te3, and bismuth telluride selenide (Bi2(Te1-xSex)3) (0 < x < 1), bismuth antimony telluride ((Bi1-ySby)2Te3) (0 < y < 1), Phys. Status Solidi B 84 (1977) 619-628. https://doi.org/10.1002/pssb.2220840226

[79] H. Haris, S.W. Harun, A.R. Muhammad, C. L. Anyi, S. J. Tan, F. Ahmad, R. M. Nor, N.R. Zulkepely, H. Arof, Passively Q-switched Erbium-doped and Ytterbium-doped fiber lasers with topological insulator bismuth selenide (Bi2Se3) as saturable absorber, Optics & Laser Technology 88 (2017) 121-127 https://doi.org/10.1016/j.optlastec.2016.09.015

[80] Y. Cheng, J. Peng, B. Xu, H. Xu, Z. Cai, J. Weng, Passive Q-switching of Pr:LiYF4 orange laser at 604 nm using topological insulators Bi2Se3 as saturable absorber, Optics & Laser Technology 88 (2017) 275-279. https://doi.org/10.1016/j.optlastec.2016.09.026

Topological Insulators: Materials and Applications
Materials Research Foundations 154 (2024) 147-171

Materials Research Forum LLC
https://doi.org/10.21741/9781644902851-8

Chapter 8

Topological Insulators for Mode-Locked Fiber Lasers

Sabahat Urossha[1] and S. S. Ali[1*]

[1]School of Physical Sciences, University of the Punjab, Lahore 54590, Pakistan

* shahbaz.sps@pu.edu.pk

Abstract

In this chapter, topological insulator materials such as graphene and bismuth telluride (Bi_2Te_3) as saturable absorbers (SA) are discussed experimentally in erbium and ytterbium-doped mode-lock fiber lasers. Ultra-short pulses at various wavelengths can be produced by modifying the crystal structures of topological insulators (TIs). This chapter provides a detailed explanation of how TIs can be created and incorporated as effective passive saturable absorbers with different fiber mode-lock lasers capable of providing basic to high-harmonic pulse production. The function of mode-locking in Er and Yb-doped fiber lasers is described experimentally by making use of various fabrication methods and optical characterizations are also discussed. The findings show that fiber-laser saturable absorbers comprised of Bi_2Te_3 and graphene have the potential to be utilized in powerful mode-locked fiber laser technologies.

Keywords

Topological Insulators, Mode-Locking, Saturable Absorber, Multi-Pulses, Optical Characterization

Contents

1.   Introduction

Mode-locked fiber lasers offer extremely efficient systems of ultra-short pulses [1]. Mode-locked fiber lasers that operate passively have drawn a lot of interest as a practical instrument for investigating multi-soliton nonlinearities [2-3]. These lasers could also be utilized to build ultra-short pulse devices that are portable, resilient, and adaptable. Realistic saturable absorbers (SAs), non-linear amplification loop mirrors, and different polarization propagation can all be used to accomplish the passive mode-locked fiber laser. Ultra-wideband fiber lasers are very significant because of the broad spectrum of uses they have, including in biomedical, material fabrication, and fundamental sciences. There are several different approaches to getting pulsed mode-locked functioning in a fiber laser. The utilization of semiconductor saturable absorber mirrors (SESAMs) is currently the most widespread one [4]. Since a realistic saturable absorber is not affected by the polarization of the resonant cavity so it is anticipated to be a very effective method of producing mode-locked pulses than the other two methods. Various SAs like SESAM [5], single-walled carbon nanotubes (SWNT), and graphene analogs are being employed in mode-lock fiber lasers. Emerging Dirac materials, Topological insulators (TI's), that display Dirac-like linear range distribution, have gained a considerable deal of focus on the realm of passive mode locking in fiber laser [6-8].

Currently, a diverse field of ultrafast science is developing enormously. Before the turn of the twenty-first century, passive mode-locked lasers were mostly used since they could generate higher peak output and femtosecond width. Even though, the laser world is moving towards ultrafast fiber lasers as a consequence of the deployment of higher flow semiconductor lasers upon the absorption of active fibers peak and because of scientific breakthroughs to novel nanostructured material. Due to their distinctive optical characteristics, 2-Dimensional nanomaterials are employed as saturable absorbers, also named mode-locker to produce extremely short pulses [9].

A series of carbon materials, comprising graphene and single-wall Nano-tubes, are suitable for fiber laser mode-locking due to their benefits of broad operating wavelength ranges as well as low expense [10, 11]. But beyond this, scientists are continually seeking improved saturable absorbers to cope with mode-locking. Topological insulators including Bi_2Se_3, Bi_2Te_3, and Sb_2Te_3 have lately gained considerable interest because of their potential usage in fiber lasers [12, 13]. Topological insulators are a novel type of quantum electronics material that displays metallurgical states on the exterior but insulating states on the inside. Because Bi_2Se_3 seems to have a topologically non-trivial band gap of roughly 0.2-0.3 eV, it exhibits an absorption peak whenever the optical wavelength is less compared to 4.1 μm (0.3 eV) [14].

Almost all mode-locked lasers are distinguished for their consistency and portability. The minimal fiber distance required for merging and doping an operational fiber, however, limits the pulse of a fiber laser's speed to many hundreds of Megahertz [15]. Several specialized uses, such as optical communication systems, synchronization, and optical sampling, necessitate extremely greater repeated frequencies. Precisely three materials have been displayed to operate in mode-locked behavior: Bismuth Selenide (Bi_2Se_3), Bismuth telluride (Bi_2Te_3), and Antimony telluride (Sb_2Te_3) [16].

A common way for obtaining high-quality pulses is to incorporate SA within a fiber laser cavity. The use of tapering and D-shaped fibers was justified by the reduced evanescent wave strength associated with saturable absorber sand and increased asymmetrical interaction distance between the guiding light and SA. Meanwhile, SA relied on D-shaped fiber, which had the drawback of requiring polarization and being challenging to manufacture. Since symmetric mode widening in the tapering waist is provided by tapered fiber, it is suggested as a straightforward, cost-effective approach. When the material is directly sprayed onto the tapered layer to produce a fiber-taper saturable absorber, it leads to significant drawbacks due to its extensive effects [17, 18].

This chapter also provides a mode-locked fiber laser demonstration with powerful Erbium (Er) and Ytterbium (Yb) doped laser, along with a topological insulator serving as SA. To the utmost of our understanding, this was one of the earliest reports of a topological insulator saturable absorber 'TISA' based fiber laser functioning at more than 1.6 μm. Furthermore, it has the greatest mean power output for passively mode-locked fiber laser relying on TISA that has been recorded by that time [19].

2. Topological insulator saturable absorber based fiber lasers

Fig. 1 provides an explanation of the topological insulators' saturable absorption. In topological insulators, either insulator bulk levels or conductive surface layers together are available. Whenever the intra-cavity strength exceeds the limit at the start of a laser, a saturable absorber begins to function.

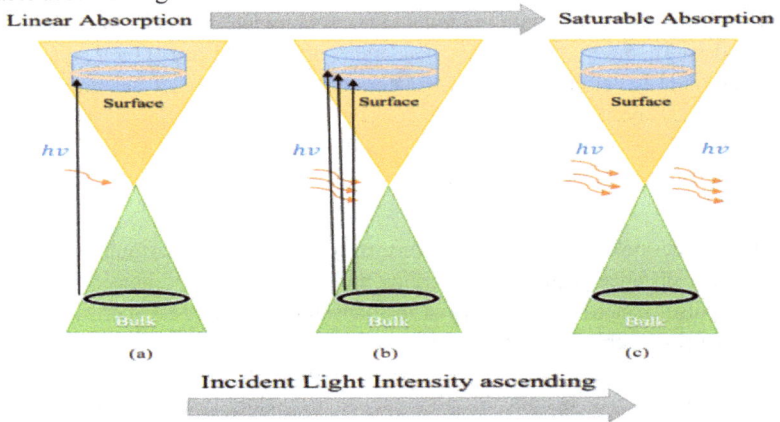

Figure 1. Process of saturable absorption in Topological insulator saturable absorber. a) Direct absorption. b) Bleaching state. c) Saturable absorption and transmission of light.

The ground states become empty as a result of all the associated electrons located in the valence band of the bulk and surface levels being excited to the conduction band by the sound pulses of lower intensities striking the SA (Fig. 1b). Saturable absorption occurs when TISA reaches this point, where it is unable to absorb any more intra-cavity photon and turns into transparency, enabling all arriving photons to flow right through (Fig. 1c). This process causes the loudest sound to ultimately break free from the saturable absorber and obtain advantage over the round trips as a result of the stimulated emission processes. Hence, the saturable absorber is always made smaller by absorbing the wings of the loudest sound. The in-phase longitudinal cavities mode constructively interferes to produce elevated output pulses. The resultant pulses get smaller, its maximum output power increases and the associated frequency spectrum is wider whereas multiple modes are in phase. In this case, TISA performs the function of a gate, opening whenever higher intensity pulse passes through it and blocking other lower-intensity (sound) pulses. TI's were intensively studied because they contain surface band structures like graphene [20].

Based on multiple types of research, TIs exhibit higher absorption rates as well as superior modulating depth over graphene for particular wavelengths [20-22].

Fabrication and characterization of TISA

Since 2012, there have been ongoing studies on the topological insulator bases upon saturable absorber 'TISA' as well as its utilization in fiber laser sources TISA is applicable to a combination of the resonant cavity and linear fiber lasers topologies.

Fabrication method

Fig. 2 provides a summary of a few established fabrication techniques. These techniques include mechanical exfoliation and liquid phase exfoliation which happen to be merely physical processes that do not include any chemical interactions, although another one is chemical vapor deposition, which is a chemical process. Even between those, three techniques that are frequently employed for laser purposes are introduced here. These are mechanical exfoliation (ME), Liquid-phase exfoliation (LPE), and chemical vapor deposition (CVD).

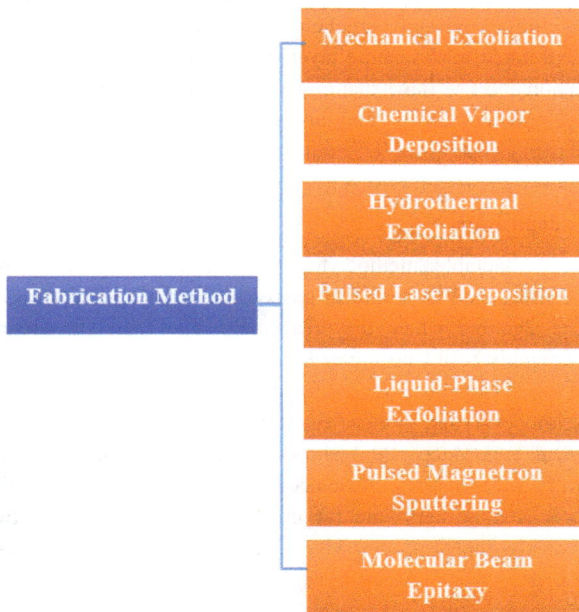

Figure 2. Classification of fabrication methods for TISA.

The major optical properties are affected by the sort of fabrication technique selected for a specific TISA. The 3 exfoliation methods, ME, HTE, and LPE, are inconsistent and sensitive to outside factors like contaminants, stuck states, and so forth. As a result, there is a wide range of variations in optical characteristics such as modulating depth, pulse duration, etc. As a consequence, it is challenging to forecast how well TISA made using these methods will function. Different fabrication techniques, like PMS, CVD, and PLD, can produce topological insulator materials in a contained way, leading to improved TI thin film purity [23, 24].

Mechanical exfoliation

Materials that are bulky are created when multiple layers of two-dimensional mono-layer material come into contact with the help of the Van der Waals force. Conversely, by resisting that force, layers can be separated from bulky substances to create high-quality two-dimensional mono- and multiple-layer compounds. An approach called mechanical exfoliation uses adhesive tape to continuously peel and disassemble bulk compounds in order to produce mono- or few-layer compounds. Gem and Novoselov in 2004, had been initially employed in the finding of graphene using graphite particles [25]. The materials that have been exfoliated into a single layer or a number of layers possess an elevated degree of completion and few flaws that make them suitable for crucial studies in science. For instance, topological insulators may also be produced using it to produce additional two-dimensional compounds [26-27]. Though the final product of this process is extremely poor, and repeating operations are needed.

Chemical vapor deposition

A significant and adaptable method for creating enormous-scale two-dimensional substances is the use of CVD. Usually, gaseous or powdered reacting substances are delivered into the chamber where they react, and with the right circumstances, two-dimensional compounds could be created from a particular chemical reaction. Direct growth of two-dimensional substances is possible on selected substrates by putting them inside the reaction chamber. When compared with solution-based manufacturing processes, chemical vapor deposition allows for more precise manipulation over the multiple layers of two-dimensional substances created by changing conditions of reaction [28]. This process is a primary one for creating readily accessible two-dimensional materials due to its significant productivity. Despite this, costs and process complexities can be substantial.

Other methods

Numerous fabrication techniques have been proposed by experts for particular two-dimensional nanomaterials. The drawback of molecular beam epitaxy (MBE) is the need for costly instruments, which could hinder it from producing outstanding topological insulator films upon large substrates [29]. Topological insulator nanostructures have also been created using the pulse magnetron sputtering technique [30, 31]. Using a laser with a pulse to serve as an ignition source, PLD evaporates materials in a vacuum chamber before allowing molecules to be deposited upon the outer layer of an appropriate substrate [32, 33]. TMDs can be created via the hydrothermal process, which uses a hydrothermal interaction of reacting substances to form crystalline nanoparticles [34]. When using a laser thinning process, heat generated through light absorption evaporates the top layers of the solid bulky compound, leaving only the lowest layer onto the substrate [35].

2.1   TISA in Erbium-doped fiber laser

The graphene topological insulator is utilized serving as SA to illustrate the creation of ultra-short pulses. Lower non-saturable setbacks and strong modulating depths (6%), were both made possible by the graphene layering on a side-polished fiber [36]. The laser could produce 280 femtosecond short pulses, which are the lowest pulsing ever produced by a soliton laser with a topological insulator based upon a saturable absorber.

Because of its distinctive optical characteristics, graphene has attracted a great deal of attention from researchers working within the discipline of ultra-fast laser. According to the pump-probe assessments, graphene exhibits a rapid relaxing transition within 70–150 femtosecond intervals, subsequently undergoing a more gradual relaxing process that occurs in the 0.5–2.0 ps duration [37, 38]. The methods were additionally employed to measure the characteristics of graphene's saturable absorption. In general, just one piece of graphene absorbed 2.3% of incoming light with a lower intensity throughout a fairly wide range of wavelengths [39]. Whenever incoming intensity is raised, a portion of this absorption may become saturating.

Graphene that is appropriate for fiber laser mode-locking may be generated in a number of ways. Among the more frequent ones are ME, LPE, and epitaxial growth through CVD.

Experimental setup

Economically accessible graphite (SGL Group) was used to create the multi-layer atomic graphene. A razor-sharp blade carefully ripped the graphite block. Pieces of graphite were produced as a consequence of this procedure. The particles were subsequently positioned on a piece of adhesive tape as well as continuously squeezed to create atomic graphene with multiple layers. After that, graphene strips were easily applied to a common fiber interface by simply pushing the ferrule against the tape's inside surface. At the side of the fiber, a graphene layer develops as a result of a powerful contact between graphite and

SiO_2. Thus, a connector is then inserted into the laser cavities and then coupled with a fresh one using an adaptor [40].

In Fig. 3, the laser configuration is displayed. It consists of a graphene-based saturable absorber positioned among 2 FC/APC interfaces, a graphene-based in-line fiber polarizing system, a 1 m in length strongly doped erbium fiber, fiber isolator, a 980/1550 single-mode wavelength division multiplexer (WDM) connector, and 10% output adapter.

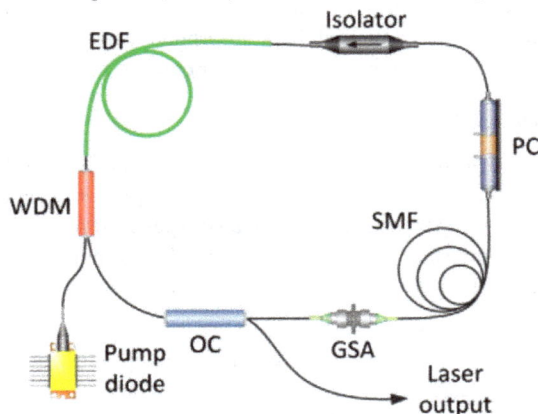

Figure 3. Setup of Mode-locked laser constructed from graphene reuse from [46]

A 980 nm laser diode is used to confront-directionally pump a laser. Cavity exclusively uses Er-doped Fiber (EDF) and single-mode fiber (SMF), both of which have opposing group velocity dispersion (GVD). Thus, the single-mode fiber length requires to be optimized in order to balance the dispersion. In this situation, the overall cavity length is around 11 m for the SMF. Soliton mode should be the laser's operating mode. The optical spectra detector, 350 MHz digitized oscilloscope, 7 GHz RF spectral detector combined with a 30 GHz photodetector [41, 42], and autocorrelation were used to monitor and assess the laser efficiency.

2.2   TISA in Ytterbium-doped fiber laser

On the basis of topological insulator: Bi_2Te_3-filled PCF (TI-PCF) saturable absorber system, a sustained evanescent pulse mode-lock function has been attained in ytterbium-doped fiber (YDF) laser oscillators.

In this section, the Bi_2Te_3 layer is placed upon the tapered fiber to create an entirely novel kind of SA using the PLD approach, which has advantages over the predecessor material in terms of reduced depositing temperature and unchanging composition. This method has an exceptional capacity to prevent the deposited Bi_2Te_3 layer from sticking to the surface of the tapered fiber.

Experimental setup

The schematic representation of ytterbium-doped fiber lasers is depicted in Fig. 4. The WDM, YDF (active media of fiber), optical coupler (OC), PI-ISO "polarization independent isolator", a fiber-tapered topological insulator saturable absorber, polarization controller (PC), SMF (20 m single-mode fiber) as well as a filter with fiber pigtails are components of ring cavities.

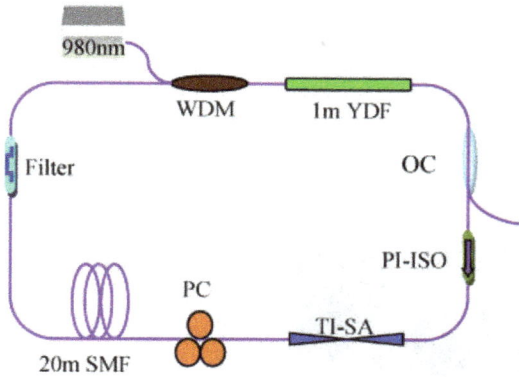

Figure 4. A schematic illustration of ytterbium doped fiber laser. Reuse from [44].

The gain medium of Yb-doped fiber having 1m length, has a 250 dB/m 980 nm absorption coefficient and a 980 nm LD (Laser diode) has been employed like a pumping root. A 10% part of the laser's strength has been disconnected from a cavity utilizing an OC. Utilizing the PI-ISO, the fiber ring cavity is made to operate in a unidirectional way. The PLD approach is used to create the fiber-taper Bi_2Te_3 saturable absorber as in Ref. [43]. The tapered fiber has a waist diameter of 55 µm and an entire length of 0.6 mm. The Bi_2Te_3 coating has been applied to the whole waist area. The Bi_2Te_3 fiber taper has been determined to have a 10% modulating depth. The PC functions by exerting pressure using a movable clamp. A birefringence develops inside the center of the fiber as a result of stress placed on the fiber. It could put a fiber laser in a correct operating condition along with sufficient nonlinear effect, making the mode-locking function simple. A spectrum filter is required for sustained mode-locking function because the fiber cavity in this instance has simply typical dispersion. As an outcome, a fiber-pigtailed filter having a bandwidth of 8 nm and a center wavelength of 1064 nm is placed inside cavities. The cavity is lengthened by an additional 20 m SMF. The fiber laser cavities measure 33.4 m in length overall. The power meter, optical spectra analyzers (YOKOGAWA AQ6370D), digital oscilloscope (Lecroy 8600A), and customized 2.5-GHz photodiode sensor are used to evaluate the fiber laser's output properties [44].

The SEM image of topological insulator Bi_2Te_3 nano-sheets is shown in Fig. 5 (a). The nanosheet's uniform hexagonal pattern could be noticed. The nano-sheets had a mean end-to-end dimension of 450 nm. The typical thickness of the sample was around 20 nm. As depicted in Fig. 5 (b), the XRD structure of topological insulator Bi_2Te_3 nano-sheets displayed a strong [006] alignment as well as certain distinguishing peaks [015 and 0015] alongside the bulk Bi_2Te_3. They diluted 0.15 mg of material inside 3 mL of deionized water and homogenized the mixture using an ultrasonic mixer for 1.5 hours to prepare the TI's liquid that was filled into the PCF. This work might have a homogenous solution that lasts for a while without subsiding [45].

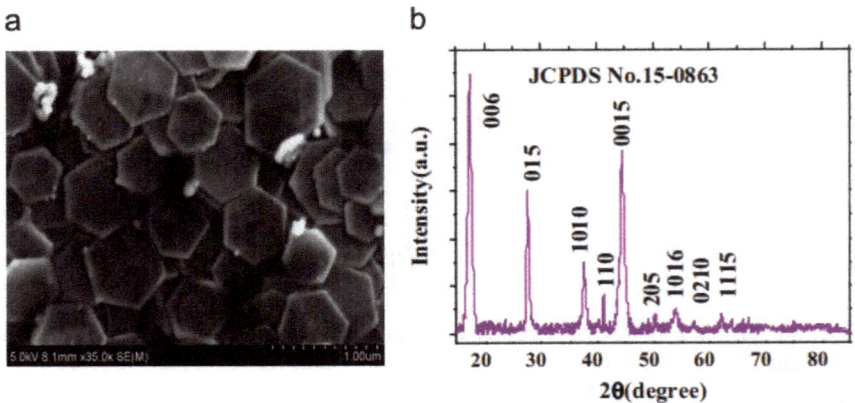

Figure 5. (a) SEM of TISA based Bi₂Te₃ Nano-films; (b) X-ray Diffraction pattern. Reuse from [49].

3.   Result and discussion

3.1   Fundamental mode-locking and optical characterization

3.1.1   Erbium-doped fiber laser

For continual wave (CW) lasing, the minimum pumping power is 30 mW. Mode-locking is possible using a standard output power of 3 mW upon boosting the pump's power to 60 mW and modifying the polarization regulator. The soliton mode-locked laser illustrated in Fig. 6 has a conventional form and distinctive Kelly's side-band that were produced at critical pumping power. On the spectrum's top, there are additionally 2 (CW) peaks that can be seen. Mode-locked laser function exhibited extremely consistent behavior and was impervious to variations in pumping power. Generally, no additional polarized regulator adjustment is required even if the pump's power is altered regardless of maintaining synchronization. To improve a radio frequency signal-to-noise ratio (SNR) as well as the optical bandwidth, a small modification of the PC's screws can be necessary.

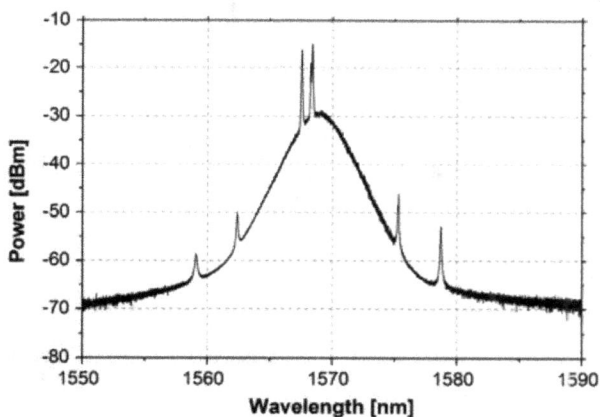

Figure 6. Resultant pulse spectra having a 0.02 nm precision at 60 mW of pumping power. Reuse from [46].

The way pumping power affected the functioning of mode-lock laser has been studied. The optical spectrum for respective pumping's powers is presented in Fig. 7(a): 200, 235, 250, 370, and 414 mW. Every output spectrum that has been measured exhibits a similar shape and lies around 1568 nm. When the pumping's power has been increased (e.g., from 3.0 to -3.0 for the highest pumping), the 3 dB bandwidth drops.

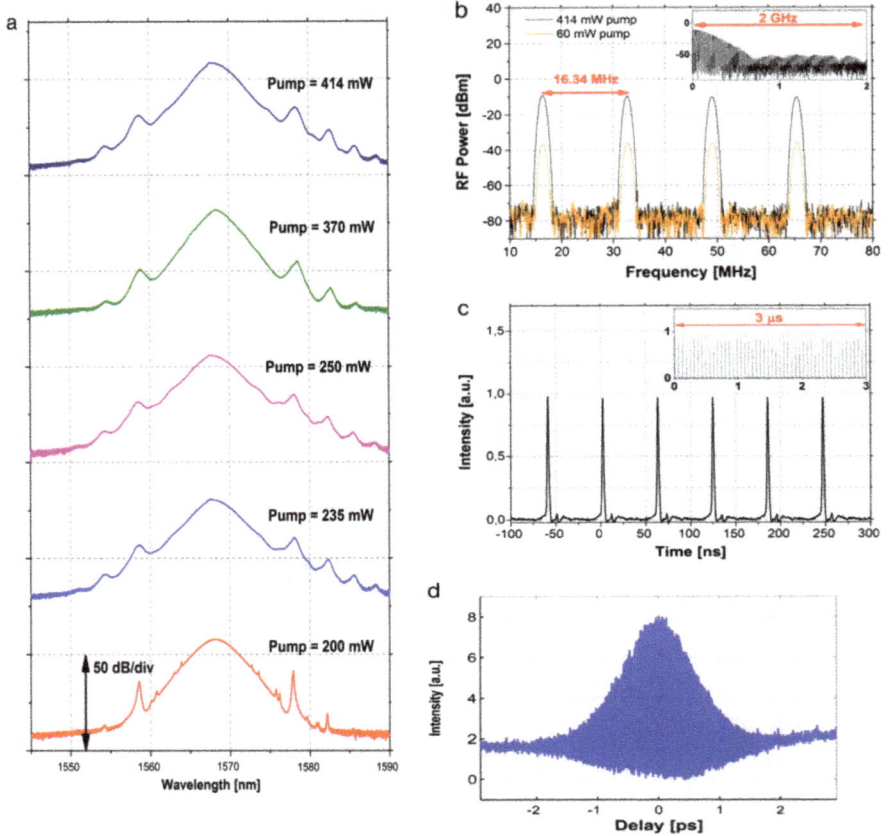

Figure 7. Effectiveness of the Mode-lock laser to the Basic Repetitive Rates: a) output pulse spectrum for various pumping power detected having 0.02 nm accuracy, b) RF spectra having 60 and 414 mW of pumping (calculated at RBW= 910 kHz). Inset: spectra in 2 GHz ranging (with 414 mW pump), c) resultant pulsing train. Inset: pulses train within 3 μs span, d) pulses autocorrelation indication at 414 mW pumping. Reuse from [46].

The determined RF spectrum over the minimum and highest pump levels is illustrated in Fig. 7(b). The repeated frequency, which had been determined to be 16.34 MHz, is equivalent to a laser's cavity that is 12 meters in length. When the power of pumping is increased, SNR is improved to around 27 dB. The resultant pulsed train is depicted in Fig. 7 (c) with pulses spaced around 61 ns apart and recorded at 414 mW pumping power that correlates to the 16.34 MHz repeating frequency. The intensity autocorrelation pattern in Fig. 7 (d) served as the foundation for calculating the pulse's period. The greatest pumping power rate was at 844 fs [46].

3.1.2 Ytterbium-doped fiber laser

For CW functioning, the laser's pump limit was 110 mW. By precisely fixing the position of the PC, the basic mode-locking function was accomplished at a pumping power of 160 mW with mean resultant power of 1.41 mW. Each cavity in this pulse pattern had a duration of 905.6 ns, as depicted in Fig. 8 (a), which corresponds to a basic repeating rate of 1.10 MHz, and an entire cavity's length that was 185 m. Fig. 8 (b) showed that the full width at half maximum (FWHM) of a single pulse profile measured 1.11 ns. As seen in Fig. 8 (c), the equivalent optical spectrum was additionally determined as well. Its 3-dB spectra bandwidth was 1.38 nm, and its center wavelength was 1064.3 nm. Additionally, the optical spectrum revealed that a shock wave oscillation in an all-normal dispersion fiber laser caused usual sharp spectral ends, demonstrating the fact that mode-locked pulses were being molded to dissipative soliton [47].

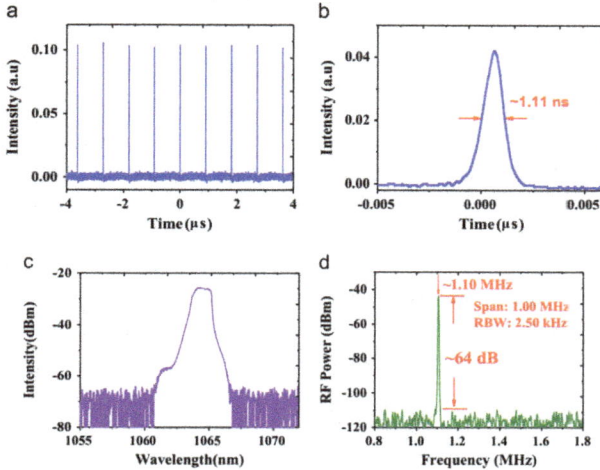

Figure 8. (a) The basic mode-locking pulsing train, (b) singular pulses width, (c) respective spectra (d) basic frequency signals and inset of respective RF spectra having range of 50 MHz. Reuse from [49].

A Combined Domains Oscilloscope (Tektronix MDO4034B-3) was used to determine associated radio frequency (RF) spectra. It was plainly visible in Fig. 8 (d) that the basic frequency had a repeated rate of 1.10 MHz and that resolution bandwidth (RBW) is measured at 2.50 kHz, having an SNR of 64 dB.

Due to maximum power-limiting impact, pulse splitting was discovered to happen by properly adjusting the PC's alignment [48]. The equivalent pumping power and mean power produced were 185 mW and 2.24 mW, correspondingly. At that pumping power, disorganized multi-pulses have been reported in which the cavity was filled with pulses at random. Two distinct pulses coexisting in the interior cavity are a common scenario, as shown in Fig. 9 (a). Unequal spacing existed between those two pulses. It was important to note that such function condition was stable and constant. By modifying the PC, a different kind of disorganized multi-pulse condition might additionally be seen, as depicted in Fig.9 (c). The cavity's interior space became fully or partially occupied by an increasing number of novel pulses. The pulses in the cavity were discovered to travel arbitrarily in relation to one another.

The related optical spectrum, which corresponds to the pulses train of Fig. 9 (a) and (d) and Fig. 9 (c) respectively, was measured in order to determine the underlying cause of the phenomenon. It was evident that the optical spectra contained CW temporal components. The development of the disorganized multi-pulses was significantly facilitated by the presence of the CW spectral constituent. A type of universal pulse's forces of interaction that connect to every single pulse may be introduced by the CW component.

The pulse's pattern might have varied group velocity and round-trip periods since the cavity's dispersing was nonzero and the pulses were positioned at various wavelengths. In a nutshell, the pulses' dispersion across the cavity got disorganized, and they were constantly moving relative to one another [49]. Furthermore, the CW component appeared as ambient sound, precisely as the findings in Ref. [49], as might be observed certainly in Fig. 9 (a)-(d). The instabilities of the pulses train got more apparent whenever the CW component's strength was increased, and as a result, the noisy background vanished from the oscilloscope mark, as seen in Fig. 9 (c). If there was no ambient noise within a cavity, it might develop as a consistent random distribution trend (as seen in Fig. 9 (a)).

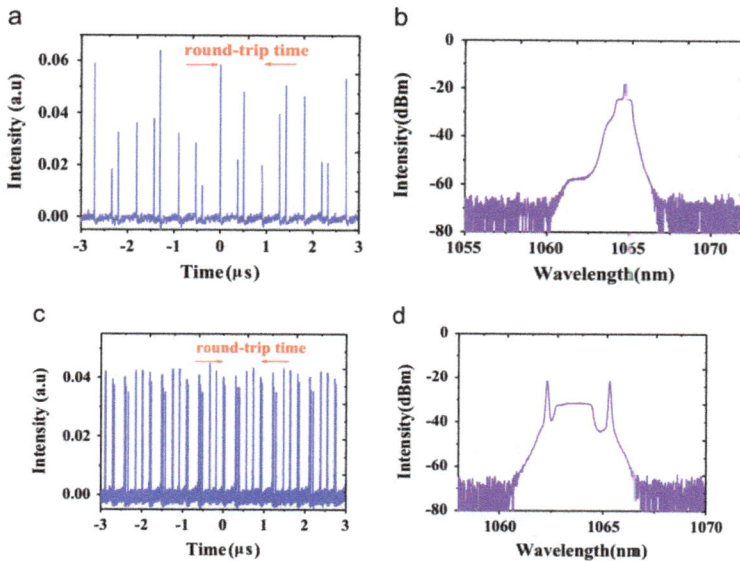

Figure 9. (a) Unordered multi-pulses oscilloscope display (b) corresponding optical spectra (c) second kind of disorganized multi-pulses (d) and respective optical spectra. Reuse from [49].

3.2 Mode-locked and Q-switched fiber lasers

3.2.1 Mode-locked fiber lasers

Mode-locked lasers produce extremely short pulses within a broad frequency band. A pulse shaper process is required in laser cavities to achieve mode locking function, i.e., the pulsed process must have a greater overall gain over continuous wave functioning in order to suppress continuous wave operation and achieve the gain contest. By placing an optical modifier in the laser cavities and deliberately imposing a periodical large transmission window in the time domain, the pulse shaper process can be dynamic. Alternatively, it can be passive by placing a saturable absorber, a passive component, into the cavities and allowing the laser to function in the pulse mode on its own. When the input optical intensity is large a saturable absorber shows lower absorption; when the intensity is insufficient, a saturable absorber displays strong absorption. The pulse peak and pulse wing suffer from lower and higher losses, respectively, during the propagation of pulses via a saturable absorber. Because of this, the pulse gets smaller through a saturable absorber each time it

Materials Research Forum LLC
https://doi.org/10.21741/9781644902851-8

passes across (saturable absorber) preceding pulsing's widening inside the cavity (caused, for example, via dispersion) balances out the pulse reducing (Kerr nonlinearity inside cavities may also participate), and a sustained pulsing function is achieved [50].

An intriguing finding is that such mode-locked laser depends upon TIs. The highest wavelengths achieved by such materials in mode-locked lasers are listed in Table-1 along with their band gap. Since their band gaps are narrow enough, there is no issue with TIs and BP. The observed operation wavelengths in lasers for TMDs, however, are substantially higher compared to the cut-off wavelength established by the band gap. This phenomenon is typically referred to as the localized stimulation or absorption of 2D materials caused by edge states that enable the absorption of photons having energies less compared to those of the typical band gaps by generating sub-band gaps with substantially lower excitation photon energy. To fully comprehend this phenomenon, more research is still needed.

Table 1. Band-gap exhibited the topological insulators' greatest operational wavelength. Reuse from [50].

TI's materials	Bandgaps	Wavelength in laser
Sb_2Te_3	Indirect for bulk: ~0.07 eV (17.7 µm) [51]	1565 nm [55]
Bi_2Se_3	Indirect for bulk: ~0.3 eV (4.14 µm) [51]	1610 nm [56]
Bi_2Te_3	Indirect for bulk: ~0.06 eV (20.7 µm) [117], ~0.165 eV (7.53 µm) if doped [51]	1935 nm [57]
MoS_2	Direct for monolayer: ~1.8 eV (690 nm) [52]	1905 nm [58]
WS_2	Direct for monolayer: ~2.1 eV (592 nm) [53]	1941 nm [59]
WSe_2	Direct for monolayer: ~1.65 eV (753 nm) [54]	1558 nm [60]
$MoSe_2$	Direct for monolayer: ~1.57 eV (791 nm) [54]	1560 nm [61]

In a nutshell, mode-locked lasers depending upon TIs had the ability to work in a broad wavelength span, obtaining wide optical spectra, and producing higher repeated rate pulse trains and extremely energetic pulses that are partly analogous to current requirements of the mode-locked lasers depending upon different instruments [50].

3.3 Q-switched fiber lasers

Another type of pulse laser is the Q-switched laser. Q-switched lasers produce broader pulses (μs-ns) and higher pulse energy in contrast to mode-locked lasers that produce ultra-short pulses (ps-fs) and a broad optical spectrum. Even so, the technique required to create a Q-switching function is remarkably identical to those of a mode-locked laser. We can activate Q-switching impact by placing an optical modifier within the cavity, or one may do it inactively by placing saturable absorbers inside the cavities. According to the passive scenario, the characteristics of the cavities and the saturable absorber determine through a cavity with a saturable absorber allows a Q-switching function or mode locking technique. The various laser operation modes have been discussed in great detail by Ref. [62].

3.4 Challenges and future perspective

Although there are numerous difficulties, the interesting outcomes achieved with TISA-based mode-locked fiber laser motivate us towards its potential uses in the future. The basic difficulty is discovering the electrical structure of topological insulator material. These materials exhibit sophisticated electrical structures because of their potent spin-orbit interaction. Thus, it is challenging to forecast theoretically the band gap energies of such materials in addition to calculating them experimentally. Analyzing the optical characteristics of topological insulators requires the use of quite complex measurement approaches. Proficiency and cutting-edge lab equipment are required for the design of the needed TISA-based mode lock fiber laser. Furthermore, contaminants, flaws, and deformation in the material can all have an impact on the bandwidth of the pulses in addition to GVD and fiber nonlinearity. Aside from that, it has been observed that band-gap values of topological insulator saturable absorbers vary with the number of multiple layers. In order to achieve the necessary TISA characteristics, the synthesis and manufacturing processes need to be improved. The consistency raises challenges when TISA is manufactured using fewer intricate methods (such ME, LPE, HTE, etc.). Imperfections will result in increased and unfavorable non-saturable losses. Presently, the literature has mainly reported single TI-based saturable absorbers for various wavelength ranges. Band gap manipulation with hetero-structure of TI material is an innovative strategy in the ultrafast field for wavelength tuning and center wavelength modification. Additionally, TISA-based ultra-short fiber lasers operating in the 2 μm range have been hardly ever described in papers. Furthermore, none or maybe very few of the manufacturing methods have used PLD, which is among the most effective methods for synthesizing crystalline multilayer materials, to create any type of TISA at 2 μm wavelength. One can anticipate a significantly shorter lifetime than the 230 fs [63] that have been recorded so far with TISA as thulium possesses a large response bandwidth (100 nm). As a result, the researchers' current ongoing aim is to extract pulses with smaller durations at 2 μm wavelength. In the coming years, it might be feasible with a regulated fabrication approach. Additionally, it had been noted from an investigation of papers that no (or maybe very few) wavelengths of fiber-based laser utilized MBE-grown TISA for mode-locking. Take into account that MBE is known as one of the more important processes for ensuring monolayer homogeneity while allowing modification of the variety of layers. However, MBE is a more expensive process that is primarily utilized in large-scale manufacturing. It is used to produce thin films atop crystalline bulk substrates. Thus, MBE-grown TISA provides an intriguing domain for research in fiber laser [23].

In the future, four factors might be taken into account to significantly enhance the laser effectiveness depending upon 2-dimensional materials. They are listed here: [50]

1. Higher speed and low noise

2. Large power and higher energy

3. Broad operation spectra

4. Spatially non-uniform polarization

Additionally, it has recently been used in the categories of gas detectors, quantum memory, and cells. Just some studies have been conducted in these fields, and those that have are essentially in the early stages. The new surface condition helps the TI-based circuits work well. A TI-based photodetector might swiftly gather the photo-generated dispersed particles due to the TIs' high surface movement in the presence of an electric field [64]. Additionally, researchers must examine the dynamical, electrical, and magnetic properties of TI's surface states. This needs to foster the growth of TI, influencing scientists all around the world.

Conclusion

In summary, a thorough exploration of mode-locked fiber lasers employing topological insulators and saturable absorbers (TISA) have been provided. The optical characteristics of widely utilized TIs are determined by quantum mechanical modeling, and it is demonstrated the manner in which this material can be employed as widespread SAs inside fiber laser cavities at various wavelength ranges. Graphics are used to explain synthesizing and production methods of topological insulator saturable absorbers. Pulsed laser deposition is discovered to be the most straightforward but effective, consistent, trustworthy, and manageable approach to creating topological insulator films over different surfaces as well as fiber placements. This study supports the notion that a novel category of nanomaterial such as TIs may act as a strong saturable absorber for mode-lock fiber lasers.

References

[1] W. Liu, L. Pang, H. Han, W. Tian, H. Chen, M. Lei, Z. Wei, 70-fs mode-locked erbium-doped fiber laser with topological insulator. Scientific reports 6 (2016) 19997. https://doi.org/10.1038/srep19997

[2] P. Grelu, J. M. S. Crespo, Multisoliton states and pulse fragmentation in a passively mode-locked fiber laser. Journal of Optics B: Quantum and Semiclassical Optics 6 (2004) S271. https://doi.org/10.1088/1464-4266/6/5/015

[3] P. Grelu, J. M. Crespo, Temporal soliton "molecules" in mode-locked lasers: Collisions, pulsations, and vibrations. Springer Berlin Heidelberg, 2008, p. 1-37. https://doi.org/10.1007/978-3-540-78217-9_6

[4] J. Boguslawski, J. Sotor, G. Sobon, J. Tarka, J. Jagiello, W. Macherzynski, K. M. Abramski, Mode-locked Er-doped fiber laser based on liquid-phase exfoliated Sb2Te3 topological insulator. Laser Physics 24 (2014) 105111. https://doi.org/10.1088/1054-660X/24/10/105111

[5] R. Paschotta, R. Häring, E. Gini, H. Melchior, U. Keller, H. L. Offerhaus, D. J. Richardson, Passively Q-switched 0.1-mJ fiber laser system at 1.53? Optics letters 24 (1999) 388-390. https://doi.org/10.1364/OL.24.000388

[6] C. Zhao, H. Zhang, X. Qi, Y. Chen, Z. Wang, S. Wen, D. Tang, Ultra-short pulse generation by a topological insulator-based saturable absorber. Applied Physics Letters 101 (2012) 211106. https://doi.org/10.1063/1.4767919

[7] Z. C. Luo, M. Liu, H. Liu, X. W. Zheng, A. P. Luo, C. J. Zhao, W. C. Xu, 2 GHz passively harmonic mode-locked fiber laser by a microfiber-based topological insulator saturable absorber. Optics letters 38 (2013) 5212-5215. https://doi.org/10.1364/OL.38.005212

[8] J. Lee, J. Koo, Y. M. Jhon, J. H. Lee, A femtosecond pulse erbium fiber laser incorporating a saturable absorber based on bulk-structured Bi2Te3 topological insulator. Optics express 22 (2014) 6165-6173. https://doi.org/10.1364/OE.22.006165

[9] H. A. Haus, Theory of mode locking with a fast saturable absorber. Journal of Applied Physics 46 (1975) 3049-3058. https://doi.org/10.1063/1.321997

[10] Z. Sun, A. G. Rozhin, F. Wang, T. Hasan, D. Popa, W. O'neill, A. C. Ferrari, A compact, high power, ultrafast laser mode-locked by carbon nanotubes. Applied Physics Letters 95 (2009) 253102. https://doi.org/10.1063/1.3275866

[11] Z. Luo, Y. Huang, J. Wang, H. Cheng, Z. Cai, C. Ye, Multiwavelength dissipative-soliton generation in Yb-fiber laser using graphene-deposited fiber-taper. IEEE Photonics Technology Letters 24 (2012) 1539-1542. https://doi.org/10.1109/LPT.2012.2208100

[12] C. Chi, J. Lee, J. Koo, J. H. Lee, All-normal-dispersion dissipative-soliton fiber laser at 1.06 μm using a bulk-structured Bi2Te3 topological insulator-deposited side-polished fiber. Laser Physics 24 (2014) 105106. https://doi.org/10.1088/1054-66/24/10/105106

[13] Z. Luo, C. Liu, Y. Huang, D. Wu, J. Wu, H. Xu, J. Weng, Topological-insulator passively Q-switched double-clad Fiber laser at 2$\mu $ m wavelength. IEEE Journal of Selected Topics in Quantum Electronics 20 (2014) 1-8. https://doi.org/10.1109/JSTQE.2014.2305834

[14] H. Yu, H. Zhang, Y. Wang, C. Zhao, B. Wang, S. Wen, J. Wang, Topological insulator as an optical modulator for pulsed solid-state lasers. Laser & Photonics Reviews 7 (2013) L77-L83. https://doi.org/10.1002/lpor.201300084

[15] J. Chen, J. W. Sickler, E. P. Ippen, F. X. Kärtner, High repetition rate, low jitter, low-intensity noise, fundamentally mode-locked 167 fs soliton Er-fiber laser. Optics letters 32 (2007) 1566-1568. https://doi.org/10.1364/OL.32.001566

[16] J. Sotor, G. Sobon, W. Macherzynski, K. M. Abramski, K. M. Harmonically mode-locked Er-doped fiber laser based on a Sb2Te3 topological insulator saturable absorber. Laser Physics Letters 11 (2014) 055102. https://doi.org/10.1088/1612-2011/11/5/055102

[17] P. Yan, A. Liu, Y. Chen, H. Chen, S. Ruan, C. Guo, G. Cao, Microfiber-based WS 2-film saturable absorber for ultra-fast photonics. Optical Materials Express 5 (2015) 479-489. https://doi.org/10.1364/OME.5.000479

[18] M. Jung, J. Koo, Y. M. Chang, P. Debnath, Y. W. Song, J. H. Lee, An all fiberized, 1.89-μm Q-switched laser employing carbon nanotube evanescent field interaction. Laser Physics Letters 9 (2012) 669. https://doi.org/10.7452/lapl.201210061

[19] Y. Meng, G. Semaan, M. Salhi, A. Niang, K. Guesmi, Z. C. Luo, F. Sanchez, High power L-band mode-locked fiber laser based on topological insulator saturable

absorber. Optics express 23 (2015) 23053-23058.
https://doi.org/10.1364/OE.23.023053

[20] D. Hsieh, D. Qian, L. Wray, Y. Xia, Y. S. Hor, R. J. Cava, M. Z. Hasan, A topological Dirac insulator in a quantum spin Hall phase. Nature 452 (2008) 970-974. https://doi.org/10.1038/nature06843

[21] A. Martinez, Z. Sun, Nanotube and graphene saturable absorbers for fiber lasers. Nature Photonics 7 (2013) 842-845. https://doi.org/10.1038/nphoton.2013.304

[22] P. Yan, R. Lin, S. Ruan, A. Liu, H. Chen, Y. Zheng, J. Hu, A practical topological insulator saturable absorber for a mode-locked fiber laser. Scientific reports 5 (2015) 1-5. https://doi.org/10.1038/srep08690

[23] S. Mondal, R. Ganguly, K. Mondal, Topological Insulators: An In-depth review of their use in modelocked fiber lasers. Annalen der Physik 533 (2021) 2000564. https://doi.org/10.1002/andp.202000564

[24] K. Wu, B. Chen, X. Zhang, S. Zhang, C. Guo, C. Li, J. Chen, High-performance mode-locked and Q-switched fiber lasers based on novel 2D materials of topological insulators, transition metal dichalcogenides, and black phosphorus: review and perspective. Optics Communications 406 (2018) 214-229. https://doi.org/10.1016/j.optcom.2017.02.024

[25] K. S. Novoselov, A. K. Geim, S. V. Morozov, D. E. Jiang, Y. Zhang, S. V. Dubonos, A. A. Firsov, Electric field effect in atomically thin carbon films. Science 306 (2004) 666-669. https://doi.org/10.1126/science.1102896

[26] J. Lee, J. Koo, Y. M. Jhon, J. H. Lee, A femtosecond pulse erbium fiber laser incorporating a saturable absorber based on bulk-structured Bi2Te3 topological insulator. Optics express 22 (2014) 6165-6173. https://doi.org/10.1364/OE.22.006165

[27] J. Sotor, G. Sobon, K. M. Abramski, Sub-130 fs mode-locked Er-doped fiber laser based on topological insulator. Optics express 22 (2014) 13244-13249. https://doi.org/10.1364/OE.22.013244

[28] H. Xia, H. Li, C. Lan, C. Li, X. Zhang, S. Zhang, Y. Liu, Ultrafast erbium-doped fiber laser mode-locked by a CVD-grown molybdenum disulfide (MoS2) saturable absorber. Optics expres 22 (2014) 17341-17348. https://doi.org/10.1364/OE.22.017341

[29] P. Yan, R. Lin, S. Ruan, A. Liu, H. Chen, Y. Zheng, J. Hu, A practical topological insulator saturable absorber for a mode-locked fiber laser. Scientific reports 5 (2015) 1-5. https://doi.org/10.1038/srep08690

[30] J. Boguslawski, G. Sobon, R. Zybala, J. Sotor, Dissipative soliton generation in Er-doped fiber laser mode-locked by Sb2Te3 topological insulator. Optics letters 40 (2015) 2786-2789. https://doi.org/10.1364/OL.40.002786

[31] M. Kowalczyk, J. Boguslawski, D. Stachowiak, J. Tarka, R. Zybala, K. Mars, K. M. Abramski, All-normal dispersion Yb-doped fiber laser mode-locked by Sb2Te3 topological insulator. SPIE. 9893 (2016) 127-133. https://doi.org/10.1117/12.2225893

[32] B. Guo, Y. Yao, P. G. Yan, K. Xu, J. J. Liu, S. G. Wang, Y. Li, Dual-wavelength soliton mode-locked fiber laser with a WS2-based fiber taper. IEEE Photonics Technology Letters 28 (2015) 323-326. https://doi.org/10.1109/LPT.2015.2495330

[33] H. Chen, I. L. Li, S. Ruan, T. Guo, P. Yan, Fiber-integrated tungsten disulfide saturable absorber (mirror) for pulsed fiber lasers. Optical Engineering 55 (2016) 081318-081318. https://doi.org/10.1117/1.OE.55.8.081318

[34] Y. Zhan, L. Wang, J. Y. Wang, H. W. Li, Z. H. Yu, Yb: YAG thin disk laser passively Q-switched by a hydro-thermal grown molybdenum disulfide saturable absorber. Laser Physics 25 (2015) 025901. https://doi.org/10.1088/1054-660X/25/2/025901

[35] A. C. Gomez, M. Barkelid, A. M. Goossens, V. E. Calado, H. S. van der Zant, G. A. Steele, Laser-thinning of MoS2: On-demand generation of a single-layer semiconductor. Nano letters 12 (2012) 3187-3192. https://doi.org/10.1021/nl301164v

[36] J. Sotor, G. Sobon, K. Grodecki, K. M. Abramski, Mode-locked Erbium-doped fiber laser based on evanescent field interaction with Sb2Te3 topological insulator. Applied Physics Letters 104 (2014) 251112. https://doi.org/10.1063/1.4885371

[37] R. W. Newson, J. Dean, B. Schmidt, H. M. van Driel, Ultrafast carrier kinetics in exfoliated graphene and thin graphite films. Optics express 17 (2009) 2326-2333. https://doi.org/10.1364/OE.17.002326

[38] J. M. Dawlaty, S. Shivaraman, M. Chandrashekhar, F. Rana, M. G. Spencer, Measurement of ultrafast carrier dynamics in epitaxial graphene. Applied Physics Letters 92 (2008) 042116. https://doi.org/10.1063/1.2837539

[39] J. Sarkar, D. K. Das, Enhanced strength in novel nanocomposites prepared by reinforcing graphene in red soil and fly ash bricks. International Journal of Minerals, Metallurgy, and Materials 26 (2019) 1322-1328. https://doi.org/10.1007/s12613-019-1835-4

[40] G. C. Sosso, S. Caravati, M. Bernasconi, Vibrational properties of crystalline Sb2Te3 from first principles. Journal of Physics: Condensed Matter 21 (2009) 095410. https://doi.org/10.1088/0953-8984/21/9/095410

[41] G. C. Sosso, S. Caravati, M. Bernasconi, Vibrational properties of crystalline Sb2Te3 from first principles. Journal of Physics: Condensed Matter 21 (2009) 095410. https://doi.org/10.1088/0953-8984/21/9/095410

[42] K. M. F. Shahil, M. Z. Hossain, V. Goyal, A. A. Balandin, Micro-Raman spectroscopy of mechanically exfoliated few-quintuple layers of Bi2Te3, Bi2Se3, and Sb2Te3 materials. Journal of Applied Physics 111 (2012) 054305. https://doi.org/10.1063/1.3690913

[43] L. Li, P. G. Yan, Y. G. Wang, L. N. Duan, H. Sun, J. H. Si, Yb-doped passively mode-locked fiber laser with Bi2Te3-deposited. Chinese Physics B 24 (2015) 124204. https://doi.org/10.1088/1674-1056/24/12/124204

[44] L. Li, Y. Wang, X. Wang, T. Lin, H. Sun, High energy mode-locked Yb-doped fiber laser with Bi2Te3 deposited on tapered fiber. Optik 142 (2007) 470-474. https://doi.org/10.1016/j.ijleo.2017.06.029

[45] P. Yan, R. Lin, H. Chen, H. Zhang, A. Liu, H. Yang, S. Ruan, Topological insulator solution filled in photonic crystal fiber for a passive mode-locked fiber laser. IEEE Photonics Technology Letters 27 (2014) 264-267. https://doi.org/10.1109/LPT.2014.2361915

[46] J. Sotor, G. Sobon, K. Krzempek, K. M. Abramski, Fundamental and harmonic mode-locking in erbium-doped fiber laser based on graphene saturable absorber. Optics communications 285 (2012) 3174-3178. https://doi.org/10.1016/j.optcom.2012.03.002

[47] C. Lecaplain, J. M. Crespo, P. Grelu, C. Conti, Dissipative shock waves in all-normal-dispersion mode-locked fiber lasers. Optics Letters 39 (2014) 263-266. https://doi.org/10.1364/OL.39.000263

[48] B. A. Malomed, Bound solitons in the nonlinear Schrödinger/Ginzburg-Landau equation. Springer Berlin Heidelberg 392 (2005) 288-294. https://doi.org/10.1007/3-540-54899-8_48

[49] Y. Peiguang, L. Rongyong, Z. Han, W. Zhiteng, C. Han, R. Shuangchen, Multi-pulses dynamic patterns in a topological insulator mode-locked ytterbium-doped fiber laser. Optics Communications 335 (2015) 65-72. https://doi.org/10.1016/j.optcom.2014.09.009

Materials Research Forum LLC
https://doi.org/10.21741/9781644902851-8

[50] K. Wu, B. Chen, X. Zhang, S. Zhang, C. Guo, C. Li, J. Chen, High-performance mode-locked and Q-switched fiber lasers based on novel 2D materials of topological insulators, transition metal dichalcogenides and black phosphorus: Review and perspective. Optics Communications 406 (2018) 214-229. https://doi.org/10.1016/j.optcom.2017.02.024

[51] J. Li, H. Luo, B. Zhai, R. Lu, Z. Guo, H. Zhang, Y. Liu, Black phosphorus: A two-dimension saturable absorption material for mid-infrared Q-switched and mode-locked fiber lasers. Scientific reports 6 (2016) 30361. https://doi.org/10.1038/srep30361

[52] Y. L. Chen, J. G. Analytis, J. H. Chu, Z. K. Liu, S. K. Mo, X. L. Qi, Z. X. Shen, Experimental realization of a three-dimensional topological insulator, Bi2Te3. Science 325 (2009) 178-181. https://doi.org/10.1126/science.1173034

[53] R. I. Woodward, E. J. R. Kelleher, R. C. T. Howe, G. Hu, F. Torrisi, T. Hasan, J. R. Taylor, Tunable Q-switched fiber laser based on saturable edge-state absorption in few-layer molybdenum disulfide (MoS2). Optics express 22 (2014) 31113-31122. https://doi.org/10.1364/OE.22.031113

[54] Y. Sun, Y. Bai, D. Li, L. Hou, B. Bai, Y. Gong, J. Bai, 946 nm Nd: YAG double Q-switched laser based on monolayer WSe2 saturable absorbers. Optics Express 25 (2017) 21037-21048. https://doi.org/10.1364/OE.25.021037

[55] J. Sotor, G. Sobon, K. M. Abramski, Sub-130 fs mode-locked Er-doped fiber laser based on topological insulator. Optics express 22 (2014) 13244-13249. https://doi.org/10.1364/OE.22.013244

[56] G. Semaan, Y. Meng, M. Salhi, A. Niang, K. Guesmi, Z. C. Luo, F. Sanchez, High power passive mode-locked L-band fiber laser based on microfiber topological insulator saturable absorber. SPIE 9893 (2016) 120-126. https://doi.org/10.1117/12.2224945

[57] JM. ung, J. Lee, J. Koo, J. Park, Y. W. Song, K. Lee, J. H. Lee, A femtosecond pulse fiber laser at 1935 nm using a bulk-structured Bi2Te3 topological insulator. Optics express 22 (2014) 7865-7874. https://doi.org/10.1364/OE.22.007865

[58] Z. Tian, K. Wu, L. Kong, N. Yang, Y. Wang, R. Chen, Y. Tang, Mode-locked thulium fiber laser with MoS2. Laser Physics Letters 12 (2015) 065104. https://doi.org/10.1088/1612-2011/12/6/065104

[59] M. Jung, J. Lee, J. Park, J. Koo, Y. M. Jhon, J. H. Lee, Mode-locked, 1.94-μm, all-fiberized laser using WS2-based evanescent field interaction. Optics express 23 (1205) 19996-20006. https://doi.org/10.1364/OE.23.019996

[60] D. Mao, X. She, B. Du, D. Yang, W. Zhang, K. Song, J. Zhao, Erbium-doped fiber laser passively mode locked with few-layer WSe2/MoSe2 nanosheets. Scientific reports 6 (2016) 23583. https://doi.org/10.1038/srep23583

[61] H. Ahmad, S. N. Aidit, N. A. Hassan, M. F. Ismail, Z. C. Tiu, Generation of mode-locked erbium-doped fiber laser using MoSe2 as saturable absorber. Optical Engineering 55 (2016) 076115-076115. https://doi.org/10.1117/1.OE.55.7.076115

[62] C. Hönninger, R. Paschotta, F. Morier-Genoud, M. Moser, U. Keller, Q-switching stability limits of continuous-wave passive mode locking. JOSA B, 16(1999), 46-56. https://doi.org/10.1364/JOSAB.16.000046

[63] C. Chi, J. Lee, J. Koo, J. H. Lee, All-normal-dispersion dissipative-soliton fiber laser at 1.06 μm using a bulk-structured Bi2Te3 topological insulator-deposited side-polished fiber. Laser Physics 24 (2014) 105106. https://doi.org/10.1088/1054-66/24/10/105106

[64] W. Tian, W. Yu, J. Shi, Y. Wang, The property, preparation and application of topological insulators: A review. Materials 10 (2017) 814. https://doi.org/10.3390/ma10070814

Chapter 9

Fundamentals Concepts of Topological Insulators: Historical Overview and Single Crystal Growth Techniques

Tanmay Bhongade [1,#], Anupras Manwar [1,#], Kunal Kumar [1,#], Prasad Kulkarni [1,#], Ramireddy Boppella [2], Suvarna R. Bathe [3], Aniruddha Chatterjee [4], and Shravanti Joshi [1*]

[1]Functional Materials Laboratory, Department of Mechanical Engineering, G. S.Mandal's Marathwada Institute of Technology, Aurangabad 431010, Maharashtra, India

[2]Department of Mechanical Engineering, Colorado State University, Fort Collins, Colorado 80523, USA

[3]Department of Physics, Shivaji University, Kolhapur 416004, India

[4]Centre for Advanced Materials Research and Technology, Department of Plastic and Polymer Engineering, Maharashtra Institute of Technology, Aurangabad 431010, Maharashtra, India

# First authors who have made equal contributions.

*shravanti.joshi@mit.asia, shravantijoshi@gmail.com

Abstract

Topological insulating materials symbolize a novel matter in the form of quantum states distinguished by distinctive core and surfaces mainly arising from macroscopic wave functions. The bulk property of the insulator grants its band structure exotic features owing to which, the material depicts an insulating core and conducting surface. Such inherent characteristics ensure that the electrons propagate along the surface. Herein, we present an informative review of topological insulating materials focusing on the basic theories, and synthesis routes to achieve single crystal structures and material properties. At the start, a historical perspective is provided with a brief discussion on topological insulators reported to date. Thereafter, a detailed account is bestowed to understand different preparation methodologies from viewpoints of defect engineering and prototype fabrication in order to realize high-quality topological insulators for subsequent roles in customized applications in the fields of quantum computing, catalysis, optoelectronic and magnetic devices. Lastly, a future outlook and the impact of topological insulators in various fields of physics, chemistry, and engineering are furnished toward the end of the chapter.

Keywords

Quantum, Band Gap, Insulator, Symmetry, Surface States, Wave Functions

Contents

1. Introduction

The advancement in any branch of science and technology is habitually driven by findings of newer material architectures. In this context, materials bestowed with exceptional quantum features are lucrative and hold greater prominence. Among such materials, topologically insulating materials belong to an interesting category with a huge surge of concerted efforts observed from researchers around the globe [1-3]. In simple words, topologically insulating matter is defined as an insulator having a metallic border in the vicinity of a vacuum [2]. The metal borders start from topological invariants that are difficult to alter until the material holds its insulation. Such insulators deal with an empirically newer characteristic of quantum mechanics paving the way for opportunities to understand Hilbert space topology from nature's viewpoint. Additionally, as the wave functions mention their electronic states in terms of Hilbert space possessing nontrivial topology hence the matter is called a topological insulator. It is a known fact that the bulk wave functions in the field of quantum mechanics are chronicled by linear integration of orthogonally normal vectors generating a well-defined set and the area traversed by such set is known as Hilbert space. In the case of solids with a definite geometrical arrangement, the quantum number can be assumed to be the vector (k) and then the wave function is considered to be charting from the vector to many points in the Hilbert space or in its extrapolated area, thus the topology appropriately suits the electronic states in the solid. Pivoting on the mode, the Hilbert space topographic anatomy leads to nontriviality, thereby resulting in the many versions of topologically insulating matters [4]. As a result of nontrivial topographic anatomy imperative relationship with the wave functions is the existence of an interface state without gaps that stems up once the insulator is physically discontinued and accepts its fate as a regular insulator in the proximity of the vacuum. This is due to the discrete nature of nontrivial topographic anatomy demonstrating energy states with the presence of gaps. So long as the energy gap stays unlocked, its topology remains unaltered. Owing to this the gap at the interface must stay locked at all times for topology to alter into trivial at the contacts. Thus, the three and two-dimensional topologically insulating matters are analogous to surface and edge states without gaps, respectively. Such an important rule for compulsory incidence of interface states without gaps is also known as bulk-boundary association, occurring during the phase transition in topologically insulating materials [5].

Most of the exotic quantum-mechanical features illustrated by the topologically insulating matter germinate from the distinctive surface and edge states. At the moment, research related to this class of materials is based on time reversal invariant arrangement. In such cases, the nontrivial topographic anatomy is secured by time reversal equilibrium [1-3]. Herein, the surface and edges states are in attendance of Dirac distributions, and for this reason, the science of contingency of Dirac fermions comes around as appropriate. Moreover, the spin degeneracy is thrust into the Dirac fermions living inside the surface and edge states time reversal invariant of the topologically insulating matter with their spin confined to the momentum. This kind of state possesses helical spin polarization paving the way to comprehend the Majorana fermions in close existence of superconductivity [6-7]. The main aim of formulating experiments for such exotic materials is to dwell deeper into their presence as well as the characteristics of such helically spin-polarized Dirac fermions in the surface states of their topographic anatomy.

The present review aims to provide a didactic opening to one of the exotic classes of materials that is topologically insulating matter, highlighting basic concepts and materials properties. In addition, fabrication methodologies developed and applied over the last several years for topological insulators using advanced physics, chemistry, and engineering principles are highlighted. As complementary data, we have also pointed out in the brief the merits and demerits of each of the fabrication techniques. Towards the end, future scope and perspectives are mentioned that are foreseen to strengthen the knowledge, endowing the researchers and engineers with remedies for unfulfilled demands, consequently guiding the progression of novel topological insulators for varied applications in science and technology.

2.   Knowledge and learning from the past - A historical perspective

The inspiration for topological insulators started flourishing at the onset of the year 1970 when the researchers, Kosterlitz and Thouless at Brown University and the University of Washington, USA, respectively provided supporting information on the existence of superconductivity at exceptionally low temperatures in some matter in contrast to its absence at higher temperatures. Likewise, a series of exotic features in magnetic material were reported by Haldane at Princeton University, USA in 1980. Interestingly, all three researchers were bestowed with the prestigious Nobel Prize in Physics in the year 2016 owing to their exceptional exploration of states and phase transformations in topological insulators [8]. In 1985, Volkov and Pankratov reported on the first-ever three-dimensional frameworks with respect to topological insulators [9]. In addition, Pankratov in association with Pakhomov and Volkov again proposed a few interesting findings in the year 1987 [10]. The presence of two-dimensional Dirac states without any gaps was demonstrated at the band transposition connect in materials such as PbTe interfaced with SnTe and HgTe interfaced with CdTe to form hetero-contacts, respectively [9-10]. Despite the fact that the classification of the topological matter and implication of time alteration equilibrium as indicated by the year 2000, the important features were already realized from the investigations done by the year 1980. Thereafter, in 2005, Kane and Mele developed a conceptual framework based on the quantum spin Hall Effect [11]. The framework greatly simplified the knowledge of time reversal symmetry through invariant construction. In 2006, a similar hypothesis was reported by Bernevig and Zhang [12]. The academic model hinted at the presence of a one-dimensional helical periphery in the quantum wells of HgTe/CdTe heterostructures. In 2007, Molenkamp et al. empirically developed the subsistence of Dirac states and helical edges at the HgTe sandwiched CdTe heterostructures [13]. Later in the year 2007, it was prognosticated that the three-dimensional topologically insulating matter may be present in compounds concerning elements like bismuth and more specifically may form quantum insulators which were previously difficult to be abridged in the form of many versions of the Spin Hall Effect. Only in the year 2008 to 2009, it was realized that the better way to acknowledge topographic materials is not by surface conductivity, but rather through three-dimensional quantized magneto-electric principles [14-15]. It was supplemented by keeping the material in the magnetic flux, resulting in a consequence analogous to that of the theoretical axion observed in the field of particle physics [16]. The experiment was demonstrated by faculties at Johns Hopkins University and Rutgers University, USA with the help of a THz spectrophotometer, confirming that the fine structure constant induces quantization of the Faraday rotary motion [17]. Furthermore, by the year 2012, Kondo insulating matter with topographic anatomy at low temperatures was acknowledged in SmB6 [18-19]. In 2014, Bose-Hubbard's theory was confirmed through experiments displaying actual manipulation in the magnetic items by the employment of topologically insulating materials [20].

3. Synthesis routes for fabrication of topological insulators

The capability of material chemists to fabricate superior eminence single crystalline materials is an imperative feature controlling the research progress pertaining to topological insulators [5]. Such crystalline materials depict continuous 3D repeating geometrical arrangement of atomic particles. Fabrication of single-crystalline materials needs significant endeavours and time as explained by the tremendous surge in research activities when compared to polycrystals. Single crystalline materials facilitate an accurate understanding of intrinsic features in the absence of any undesired effect from defects. Moreover, the crystals also allow interpretations of any anisotropic characteristics in addition to different crystallographic axes derived from crystal composition-induced anisotropies. Such merits make single crystalline matter especially indispensable for analyzing topographic anatomy by means of different spectrophotometers and transport methodologies. Among the various crystal synthesis methods on hand to materials chemists, widely utilized techniques are the optical floating zone, metal-flux route, Czochralski method, chemical vapor deposition, and Bridgman principle (**Fig.1**). The concerned route relies on several parameters, namely dopant size and concentration, ultimate crystal structure and dimensions, stability, volatile nature and soon [21]. The next subsections examine several merits and demerits of the prominent growth techniques for the preparation of single crystals with topographic anatomy.

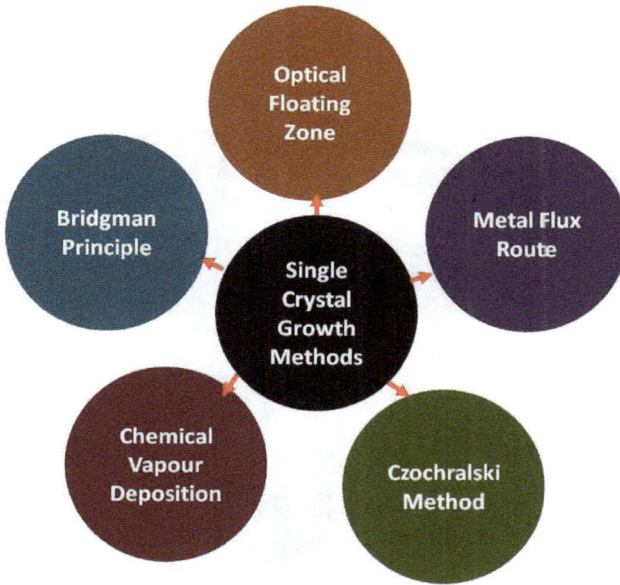

Figure. 1. Schematic representation depicts different single crystals growth techniques.

3.1 Optical floating zone

The optical floating zone is an extremely helpful route to realize single crystalline materials with larger crystals having exceptionally high melting points. In comparison to several other techniques, this route doesn't utilize a crucible for growing crystals as represented in Fig.2 [22]. The equipment employs halogen as well as xenon light sources containing 2 to 4 semi-ellipsoidal reflecting surfaces. At the foci of one lamp, each mirror is placed, whereas the other focus contains hot regions [21-23]. The assembly of 4 reflecting panels reported in the current times is equipped with laser diodes with higher capacity replacing the complete light source and reflective surface construction. While using the method, the data on several occasions reports spiky temperature contour hinting at the small molten region that ultimately results in reduced evaporation [24]. The assembly in general is effective for synthesizing single crystalline matter accommodating volatile organic compounds or dissimilar melts when compared with the furnaces fitted with light sources. The method is widely utilized for growing single crystals containing carbides (SiC, BC,

FeC, AlC, etc), silicides (CrSi, CoSi, RbSi, Cu_5Si, etc), borides (LaB_6, SiB_6, etc) and so on [21-24].

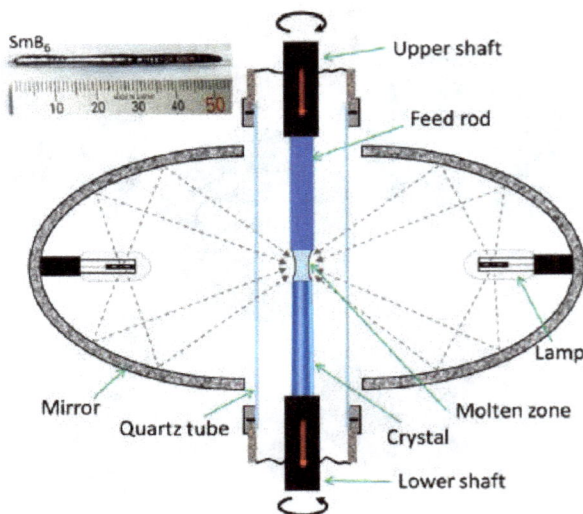

Figure.2. Schematic representation depicting the optical floating zone growth technique utilized for single crystals (Reprinted from reference no. 21 with permission under Creative Commons Attribution 4.0 International. Copyright 2020 American Chemical Society). Inset illustrates samarium hexaboride (SmB₆) crystal with topographic anatomy (Reprinted with permission from reference no. 24. Copyright 2020 Elsevier Ltd).

3.2 Metal flux route

A metal flux technique involves dissolving essential elements in an appropriate low-melting metallic flux to achieve a supersaturated medium at elevated energy [25]. The flux on supervised cooling yields single crystalline material of preferred concentration. The route is convenient and flexible to synthesize superior-grade single crystals of different constituents varying from oxides, intermetallic Heusler alloys, semimetals, chalcogenides, and so on [26-28]. A primary merit of the route is the absence of sophisticated instruments other than an apposite crucible and furnace with a standardized temperature circulation unit [21]. Popularly employed metallic fluxes are the mixtures of potassium and sodium chloride, bismuth, aluminium, tin, antimony, and indium to name a few [27-28].

3.3   Czochralski method

The Czochralski method is broadly employed in the manufacturing of different types of single crystals containing silicon and germanium at industrial sites [29-30]. Briefly, in a crucible containing molten metal, a miniature seed crystal is pressed from the above into surface of the metallic slurry [30-32]. The molten metal temperature is maintained in such a way that a small segment of the pressed seed is deliquesced. Thereafter, the seed is carefully removed through a rotary motion leading formation of newer crystals at the boundary. Normally, the newly formed crystal starts growing in a cylindrical form, wherein the diameter is conveniently adjusted by increasing melt heat, rotating, and stretching rate of the seed crystal. The major advantage of the method lies in producing superior-quality single crystals that can be used as congruent topological insulators. In recent times, materials such as PdGa, CoSi, FeSi, MnSi, and even $Fe_{1-x}Co_xSi$ have been reported [29-32]. Fig.3 illustrates the phase equilibrium diagrams developed for the palladium-gallium (PdGa) bimetallic alloy system and subsequent single crystalline intermetallic compound of PdGa grown using the Czochralski technique.

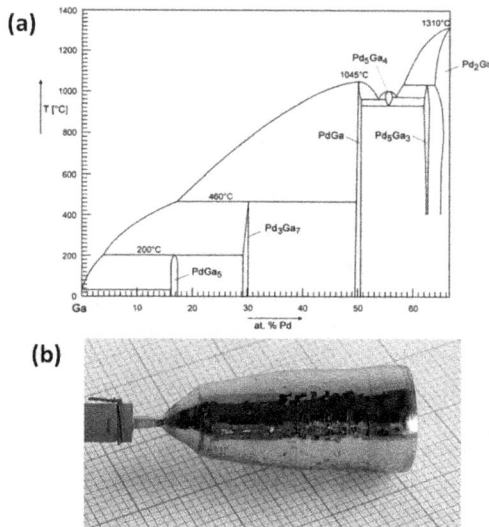

Figure.3. (a) Phase equilibrium diagrams for PdGa alloy system. (b) Single crystalline intermetallic compound of PdGa synthesized using Czochralski technique using $Pd_{45}Ga_{55}$ molten slurry (grid size: 1 mm). Reprinted with permission from reference no. 29. Copyright 2010 Elsevier Ltd.

Materials Research Forum LLC
https://doi.org/10.21741/9781644902851-9

3.4   Chemical vapour deposition

For the past few decades, chemical vapor deposition has been a popular technique for solids purification and transformation to single crystalline materials [21]. All types of solids, that is, metal, non-metal, and intermetallic having the ability to convert into a gaseous state can be used in the process. The technique involves reversible chemical reactions, wherein the vapours of the solid are turned into gaseous species in the presence of halogen compounds subsequently depositing them as single crystals on the desired substrate in various morphologies. Examples include binary and/or tertiary and/or compounds with constituent elements such as Mo, W, P, Te, Hf, Zr, Sn, and Co [33-37]. In the case of standard binary/tertiary/quaternary compounds, the combination of the constituent elements in powder form is placed in a sealed quartz container filled with halogen-based compounds (transport medium) in the vacuum. The mix is usually kept in a horizontal tube maintaining two heating zones (source end and sink end) at two ends thus creating a temperature difference that acts as a path for gaseous species during the diffusion. Herein, the crystal deposition takes place at low or high temperatures based on the chemical reactions. In many cases, a transport medium is required as the vapor pressure created as a result of chemical reactions is very small [33-35]. In this method, it is normal to utilize halogen and metal halide compounds as transport mediums. It skilfully reacts with starting precursor to vaporize it followed by diffusion and deposition. The mass transfer of the reactants in gaseous form at the far end of the horizontal tube across for deposition as single crystalline material is mediated by temperature difference created as a result of two heating zones [37].

3.5   Bridgman principle

The compounds having high thermodynamic stability require higher operating temperatures for their purification and growth into single crystals [38-39]. In such a scenario, the Bridgman principle is an excellent choice for the melting of congruent materials as well as compounds showing minimal to no phase transformation from ambient temperature to melting point [40-42]. The principle is widely used for growing huge crystals owing to quick technique and convenience. At the bottom of a furnace with a sharp conical end, the material is kept that is subjected to temperature differences created either by single or double heating zones. Thereafter, the material is supplied heat above its melting temperature by means of temperature difference. Once the heating is completed, the furnace is slowly cooled down through steady velocity to realize stable transformation. Once the temperature at the conical end goes below the solidification temperature, the seed starts the growth of the crystal at the molten melt-seed boundary. As the furnace starts to lose its heat, the complete volume of molten melt turns into a single crystalline solid ingot.

Examples include nontrivial topological insulating materials in binary and tertiary form with constituent elements consisting of Rh, Mn, Sn, Co, Ga, and V, and so on [38-42].

4. Outlook and future perspectives

The neoteric disclosures of topologically insulating matter analogous to many other findings in physics, chemistry, biology, and engineering empower newer technologies, thus enriching our knowledge for varied applications. The atypical matter consisting of the non-metallic core with a metallic boundary finds usage in quantum computing, catalysis, energy generation, spintronics, and optical, or even magneto-electric prototypes. Additionally, its amalgamation with superconductors is foreseen to result in newer versions of quantum bits. Such a state of matter continues to impact the fundamental areas in the field of condensed matter physics, confirming that the topological influences that were assumed to be limited only to lower activation energies, minimized dimensions or inflated magnetic flux can play a crucial role in understanding the physics of apparently typical macroscopic materials at standard settings.

Conclusions

The current review highlighted an academic account of topologically insulating matter focusing on fundamental concepts, various methodologies utilized for its preparation as well as principles of transport phenomenon. Firstly, historical perspectives that enabled precious learnings of past and present for future endeavours were discussed in brief. Thereafter, different routes for the synthesis of single crystals were reviewed to understand the impact of operation parameters, chemical composition, and aspect ratio on the overall features of topological insulators. Lastly, a discussion was placed to understand the way forward through impact and future perspectives. The work presented here is anticipated to broaden the interpretation to gain not only academic knowledge but also insights from viewpoints of research for the evolution of topological insulators with respect to fabrication, prototyping, and commercialization for application in multidisciplinary areas.

Acknowledgments

SJ acknowledges the Department of Science and Technology (DST), Ministry of Science & Technology, Government of India for generous funds sanctioned under Mission Innovation Carbon Capture Challenge (IC#3) with grant number - DST/TM/EWO/MI/CCUS/27/2019. SJ also acknowledges "Marathwada MedTech Lab", a Medical Device Rapid Prototyping Facility funded by the Biotechnology Industry Research Assistance Council (BIRAC), Government of India under the industry-academia collaborative mission for Accelerating Discovery Research to Early Development of Biopharmaceuticals - Innovate in India (I3) at GSM MIT Aurangabad campus.

References

[1] M. Z. Hasan, C. L. Kane, Colloquium: Topological insulators, Rev. Mod. Phys. 82 (2010) 3045. https://doi.org/10.1103/RevModPhys.82.3045

[2] J. E. Moore, The birth of topological insulators, Nature 464 (2010) 194. https://doi.org/10.1038/nature08916

[3] X.L. Qi, S.C. Zhang, Topological insulators and superconductors, Rev. Mod. Phys. 83 (2011) 1057. https://doi.org/10.1103/RevModPhys.83.1057

[4] A. P. Schnyder, S. Ryu, A. Furusaki, and A. W. W. Ludwig, Classification of topological insulators and superconductors in three spatial dimensions, Phys. Rev. B, 78 (2008) 195125. https://doi.org/10.1103/PhysRevB.78.195125

[5] Y. Ando, Topological insulator materials, J. Phys. Soc. Japan 82 (2013) 102001. https://doi.org/10.7566/JPSJ.82.102001

[6] F. Wilczek, Majorana returns, Nat. Phys. 5 (2009) 614. https://doi.org/10.1038/nphys1380

[7] L. Fu and C. L. Kane, Superconducting proximity effect and Majorana fermions at the surface of a topological insulator, Phys. Rev. Lett. 100 (2008) 096407. https://doi.org/10.1103/PhysRevLett.100.096407

[8] Neil Savage, Topology shapes a search for new materials, ACS Cent. Sci. 4 (2018) 523-526. https://doi.org/10.1021/acscentsci.8b00275

[9] B. A. Volkov, O. A. Pankratov, Two-dimensional massless electrons in an inverted contact, JETP Lett. 42 (1985) 178-181.

[10] O.A. Pankratov, S.V. Pakhomov, B.A. Volkov, Supersymmetry in heterojunctions: Band-inverting contact on the basis of Pb1-xSnxTe and Hg1-xCdxTe, Solid State Commun. 61 (1987) 93-96. https://doi.org/10.1016/0038-1098(87)90934-3

[11] C. L. Kane, E. J. Mele, Quantum spin hall effect in graphene, Phys. Rev. Lett. 95 (2005) 226801. https://doi.org/10.1103/PhysRevLett.95.226801

[12] A. B. Bernevig, S.-C. Zhang, Quantum spin Hall effect, Phys. Rev. Lett. 96 (2006) 106802. https://doi.org/10.1103/PhysRevLett.96.106802

[13] M. König, S. Wiedmann, C. Brüne, A. Roth, H. Buhmann, L. W. Molenkamp, X.L. Qi, S.C. Zhang, Quantum spin Hall insulator state in HgTe quantum wells, Science. 318 (2007) 766-770. https://doi.org/10.1126/science.1148047

[14] X.-L. Qi, T. L. Hughes, S.-C. Zhang, Topological field theory of time-reversal invariant insulators, Phys. Rev. B, 78 (2008) 195424. https://doi.org/10.1103/PhysRevB.78.195424

[15] A. M. Essin, J. E. Moore, D. Vanderbilt, Magnetoelectric polarizability and axion electrodynamics in crystalline insulators, Phys. Rev. Lett. 102 (2009) 146805. https://doi.org/10.1103/PhysRevLett.102.146805

[16] F. Wilczek, Frank, Two applications of axion electrodynamics, Phys. Rev. Lett. 58 (1987) 1799-1802. https://doi.org/10.1103/PhysRevLett.58.1799

[17] L. Wu, M. Salehi, N. Koirala, J. Moon, S. Oh, N. P. Armitage, Quantized Faraday and Kerr rotation and axion electrodynamics of a 3D topological insulator, Science. 354 (2016) 1124-1127. https://doi.org/10.1126/science.aaf5541

[18] E. S. Reich, Hopes surface for exotic insulator: Findings by three teams may solve a 40-year-old mystery, Nature: Springer Science and Business Media LLC, 492 (2012) 165. https://doi.org/10.1038/492165a

[19] M. Dzero, K. Sun, V. Galitski, P. Coleman, Topological Kondo insulators, Phys. Rev. Lett. 104 (2010) 106408. https://doi.org/10.1103/PhysRevLett.104.106408

[20] A. R. Mellnik, J. S. Lee, A. Richardella, J. L. Grab, P. J. Mintun, M. H. Fischer, A. Vaezi, A. Manchon, E. A. Kim, N. Samarth, D. C. Ralph, Spin-transfer torque generated by a topological insulator, Nature 511 (2014) 449-451. https://doi.org/10.1038/nature13534

[21] N. Kumar, S. N. Guin, K. Manna, C. Shekhar, C. Felser, Topological quantum materials from the viewpoint of chemistry, Chem. Rev. 121 (2021) 2780-2815 https://doi.org/10.1021/acs.chemrev.0c00732

[22] M. C. Hatnean, M. R. Lees, D. M. Paul, G. Balakrishnan, Large high-quality single-crystals of the new topological Kondo insulator SmB6, Sci. Rep. 3 (2013) 3071. https://doi.org/10.1038/srep03071

[23] B. S. Tan, Y. T. Hsu, B. Zeng, M. C. Hatnean, N. Harrison, Z. Zhu, M. Hartstein, M. Kiourlappou, A. Srivastava, M. D. Johannes, Unconventional Fermi surface in an insulating state, Science 349 (2015) 287. https://doi.org/10.1126/science.aaa7974

[24] Y. Kaneko, Y. Tokura, Floating zone furnace equipped with a high power LASER of 1 kW composed of five smart beams, J. Cryst. Growth 533 (2020) 125435. https://doi.org/10.1016/j.jcrysgro.2019.125435

[25] M. G. Kanatzidis, R. Pöttgen, W. Jeitschko, The metal flux: A preparative tool for the exploration of intermetallic compounds. Angew. Chem. Int. Ed. 44 (2005) 6996−7023. https://doi.org/10.1002/anie.200462170

[26] D. E. Bugaris, H. C. Loye, Materials discovery by flux crystal growth: Quaternary and higher order oxides. Angew. Chem. Int. Ed. 51 (2012) 3780−3811. https://doi.org/10.1002/anie.201102676

[27] P. C. Canfield, Z. Fisk, Growth of single crystals from metallic fluxes. Philos. Mag. B, 65 (1992) 1117−1123. https://doi.org/10.1080/13642819208215073

[28] H. Okamoto, T. Massalski, Binary alloy phase diagrams; ASM International: Materials Park, OH, 1990.

[29] P. Gille, T. Ziemer, M. Schmidt, K. Kovnir, U. Burkhardt, M. Armbrüster, Growth of large PdGa single crystals from the melt, Intermetallics 18 (2010) 1663−1668. https://doi.org/10.1016/j.intermet.2010.04.023

[30] V. A. Dyadkin, S. V. Grigoriev, D. Menzel, D. Chernyshov, V. Dmitriev, J. Schoenes, S. V. Maleyev, E. V. Moskvin, H. Eckerlebe, Control of chirality of transition-metal monosilicides by the Czochralski method. Phys. Rev. B: Condens. Matter Mater. Phys. 84 (2011) 014435. https://doi.org/10.1103/PhysRevB.84.014435

[31] N. B. M. Schröter, S. Stolz, K. Manna, F. de Juan, M. G. Vergniory, J. A. Krieger, D. Pei, T. Schmitt, P. Dudin, T. K.; Kim, Observation and control of maximal Chern numbers in a chiral topological semimetal, Science 369 (2020) 179−183. https://doi.org/10.1126/science.aaz3480

[32] P. Sessi, F. R. Fan, F. Kuster, K. Manna, N. B. M. Schroter, J. R. Ji, S. Stolz, J. A. Krieger, D. Pei, T. K. Kim, P. Dudin, C. Cacho, R. Widmer, H. Borrmann, W. Shi, K. Chang, Y. Sun, C. Felser, S. S. P. Parkin, Direct observation of handedness-dependent quasiparticle interference in the two enantiomers of topological chiral semimetal PdGa. Nat. Commun. 11 (2020) 3507. https://doi.org/10.1038/s41467-020-17261-x

[33] Z. Li, H. Chen, S. Jin, D. Gan, W. Wang, L. Guo, X. Chen, Weyl semimetal TaAs: Crystal growth, morphology, and thermodynamics. Cryst. Growth Des. 16 (2016) 1172−1175. https://doi.org/10.1021/acs.cgd.5b01758

[34] F. Arnold, C. Shekhar, S. C. Wu, Y. Sun, R. D. dos Reis, N. Kumar, M. Naumann, M. O. Ajeesh, M. Schmidt, A. G. Grushin, J. H. Bardarson, M. Baenitz, D. Sokolov,

H. Borrmann, M. Nicklas, C. Felser, E. Hassinger, B. Yan, Negative magnetoresistance without well-defined chirality in the Weyl semimetal TaP. Nat. Commun. 7 (2016)11615. https://doi.org/10.1038/ncomms11615

[35] J. Wang, P. Yox, K. Kovnir, Flux growth of phosphide and arsenide Crystals. Front. Chem. 8 (2020) 00186. https://doi.org/10.3389/fchem.2020.00186

[36] X. Huang, L. Zhao, Y. Long, P. Wang, D. Chen, Z. Yang, H. Liang, M. Xue, H. Weng, Z. Fang, Observation of the chiral-anomaly-induced negative magnetoresistance in 3d Weyl semimetal TaAs. Phys. Rev. X 5 (2015) 031023. https://doi.org/10.1103/PhysRevX.5.031023

[37] Y. Qi, P. G. Naumov, M. N. Ali, C. R. Rajamathi, W. Schnelle, O. Barkalov, M. Hanfland, S. C. Wu, C. Shekhar, Y. Sun, V. Suß, M. Schmidt, U. Schwarz, E. Pippel, P. Werner, R. Hillebrand, T. Forster, E. Kampert, S. Parkin, R. J. Cava, C. Felser, B. Yan, S. A. Medvedev, Superconductivity in Weyl semimetal candidate MoTe2. Nat. Commun. 7 (2016) 11038. https://doi.org/10.1038/ncomms11038

[38] D. Schulz, S. Ganschow, D. Klimm, K. Struve, Inductively heated Bridgman method for the growth of zinc oxide single crystals. J. Cryst. Growth 310 (2008) 1832–1835. https://doi.org/10.1016/j.jcrysgro.2007.11.050

[39] K. Hoshikawa, E. Ohba, T. Kobayashi, J. Yanagisawa, C. Miyagawa, Y. Nakamura, Growth of β-Ga2O3 single crystals using vertical Bridgman method in ambient air. J. Cryst. Growth 447 (2016) 36–41. https://doi.org/10.1016/j.jcrysgro.2016.04.022

[40] B. Jariwala, D. Voiry, A. Jindal, B. A. Chalke, R. Bapat, A. Thamizhavel, M. Chhowalla, M. Deshmukh, A. Bhattacharya, Synthesis and characterization of ReS2 and ReSe2 layered chalcogenide single crystals. Chem. Mater. 28 (2016) 3352–3359. https://doi.org/10.1021/acs.chemmater.6b00364

[41] S. Johnsen, Z. Liu, J. A. Peters, J. H. Song, S. C. Peter, C. D. Malliakas, N. K. Cho, H. Jin, A. J. Freeman, B. W. Wessels, Thallium chalcogenide-based wide-band-gap semiconductors: TlGaSe2 for radiation detectors. Chem. Mater. 23 (2011) 3120–3128. https://doi.org/10.1021/cm200946y

[42] L. D. Zhao, S. H. Lo, Y. Zhang, H. Sun, G. Tan, C. Uher, C. Wolverton, P. V. Dravid, M. G. Kanatzidis, Ultralow thermal conductivity and high thermoelectric figure of merit in SnSe crystals. Nature 508 (2014) 373–377. https://doi.org/10.1038/nature13184

Keyword Index

About the Editors

Dr. Inamuddin is working as an Assistant Professor at the Department of Applied Chemistry, Aligarh Muslim University, Aligarh, India. He obtained a Master of Science degree in Organic Chemistry from Chaudhary Charan Singh (CCS) University, Meerut, India, in 2002. He received his Master of Philosophy and Doctor of Philosophy degrees in Applied Chemistry from Aligarh Muslim University (AMU), India, in 2004 and 2007, respectively. He has extensive research experience in multidisciplinary fields of Analytical Chemistry, Materials Chemistry, and Electrochemistry and, more specifically, Renewable Energy and Environment. He has worked on different research projects as a project fellow and senior research fellow funded by the University Grants Commission (UGC), Government of India, and the Council of Scientific and Industrial Research (CSIR), Government of India. He has received the Fast Track Young Scientist Award from the Department of Science and Technology, India, to work in the area of bending actuators and artificial muscles. He has also received the Sir Syed Young Researcher of the Year Award 2020 from Aligarh Muslim University. He has completed four major research projects sanctioned by the University Grant Commission, Department of Science and Technology, Council of Scientific and Industrial Research, and Council of Science and Technology, India. He has published 210 research articles in international journals of repute and nineteen book chapters in knowledge-based book editions published by renowned international publishers. He has published 180 edited books with Springer (U.K.), Elsevier, Nova Science Publishers, Inc. (U.S.A.), CRC Press Taylor & Francis Asia Pacific, Trans Tech Publications Ltd. (Switzerland), IntechOpen Limited (U.K.), Wiley-Scrivener, (U.S.A.) and Materials Research Forum LLC (U.S.A). He is a member of various journals' editorial boards. He has served as Associate Editor for journals (Environmental Chemistry Letter, Applied Water Science and Euro-Mediterranean Journal for Environmental Integration, Springer-Nature), Frontiers Section Editor (Current Analytical Chemistry, Bentham Science Publishers), Editorial Board Member (Scientific Reports-Nature) and Review Editor (Frontiers in Chemistry, Frontiers, U.K.) He has also guest-edited various special thematic issues for the journals of Elsevier, Bentham Science Publishers, and John Wiley & Sons, Inc. He has attended as well as chaired sessions at various international and national conferences. He has worked as a Postdoctoral Fellow, leading a research team at the Creative Research Initiative Center for Bio-Artificial Muscle, Hanyang University, South Korea, in the field of renewable energy, especially biofuel cells. He has also worked as a Postdoctoral Fellow at the Center of Research Excellence in Renewable Energy, King Fahd University of Petroleum and Minerals, Saudi Arabia, in the field of polymer electrolyte membrane fuel cells and computational fluid dynamics of polymer electrolyte membrane fuel cells. He is a life member of the Journal of the Indian Chemical Society. His research interest includes ion exchange materials, a sensor for heavy metal ions, biofuel cells, supercapacitors and bending actuators.

Tariq Altalhi, PhD, is working as Associate Professor in the Department of Chemistry at Taif University, Saudi Arabia. He received his doctorate degree from University of Adelaide, Australia in the year 2014 with Dean's Commendation for Doctoral Thesis Excellence. He has worked as head of Chemistry Department at Taif university and Vice Dean of Science College. In 2015, one of his works was nominated for Green Tech awards from Germany, Europe's largest environmental and business prize, amongst top 10 entries. He has co-edited various scientific books. His group is involved in fundamental multidisciplinary research in nanomaterial synthesis and engineering, characterization, and their application in molecular separation, desalination, membrane systems, drug delivery, and biosensing. In addition, he has established key contacts with major industries in Kingdom of Saudi Arabia.

Dr. Mohammad A. Jafar Mazumder has been serving as a Professor of Chemistry at King Fahd University of Petroleum & Minerals (KFUPM), Saudi Arabia. He has extensive experience in designing, synthesizing, and characterizing various organic compounds, ionic and thermo-responsive polymers for corrosion, water treatment, and biomedical applications. Dr. Jafar Mazumder obtained his B.Sc (Hons.), M.Sc (Chemistry) from Aligarh Muslim University, India, MS (Chemistry) from KFUPM, Saudi Arabia, and Ph.D. in Chemistry (2009) from McMaster University, Canada.

In more than 20 years of academic research, Dr. Jafar Mazumder has had the opportunity to work with several international collaborative research groups and has exposed himself to a broad range of research areas. Dr. Jafar Mazumder secured 8 US patents, published more than 85 articles in peer-reviewed journals, 37 conference proceedings, 9 book chapters, and co-edited 4 books with Springers and Trans Tech publications. He is awarded as a Fellow of the Royal Society of Chemistry and Chartered Chemist, Association of Chemical Profession of Ontario, Canada. Besides, Dr. Jafar Mazumder is a member of the American Chemical Society (ACS), Canadian Society for Chemistry (CSC), Canadian Biomaterial Society (CBS), and a life member of the Bangladesh Chemical Society (BCS). In his academic career, he was awarded numerous national and international scholarships and awards including the prestigious Indian Council for Cultural Relations (ICCR) Scholarship from Govt. of India for undergraduate studies in India, Aligarh Muslim University undergraduate & graduate Gold medal, and certificate of excellence from the Ministry of Human Resource Development, Govt. of India, and

MITACS postdoctoral fellowship (Canada) for pursuing postdoctoral research in Chemical and Biomedical Engineering.

Currently, Dr. Jafar Mazumder is actively involved in several ongoing university (KFUPM), government (KACST, NSTIP), and client (Saudi Aramco) funded projects in the capacity of principal and co-investigators. His current research interest includes the design, synthesis, and characterization of various modified monomers and polymers for potential use in the inhibition of mild steel corrosion in oil and gas industries and the preparation of multilayered polyelectrolyte coated membranes for the removal of heavy metals and organic contaminants from aqueous water samples. The long-term scientific goal of Dr. Jafar Mazumder is not merely to make science fun and entertaining for people. It is to engage them with a multidisciplinary scientific mission at a deeper level to create a space through which they can interact with scientific ideas, develop connections between science, engineering, and biology, and thoughts of their own to contribute to society. He feels this goal and engaging personality make him a pleasant person to work with and help inspire his co-workers in any professional setting.